Contents

Section 1 Intro... W9-DHW-301

Unit 1   General Principles of Electric Motor Control   1
Unit 2   Fractional Horsepower Manual Motor Starters   12
Unit 3   Magnetic Line Voltage Starters   16

Section 2 Control Pilot Devices

Unit 4   Pushbutton Control   27
Unit 5   Relays and Contactors   29
Unit 6   Timing Relays   34
Unit 7   Pressure Switches and Regulators   43
Unit 8   Float Switches   44
Unit 9   Flow Switches   45
Unit 10   Limit Switches   47
Unit 11   Phase Failure Relays   48
Unit 12   Solenoid Valves   49
Unit 13   Temperature Switches   52

Section 3 Circuit Layout, Connections, and Symbols

Unit 14   Symbols   53
Unit 15   The Interpretation and Application of Simple Wiring
           and Elementary Diagrams   56

Section 4 Basic Control Circuits

Unit 16   Two-Wire Controls   67
Unit 17   Three-Wire and Separate Controls   68
Unit 18   Hand-Off Automatic Controls   70
Unit 19   Multiple Pushbutton Stations   72
Unit 20   Interlocking Methods for Reversing Control   74
Unit 21   Sequence Control   78
Unit 22   Time-Delay, Low-Voltage Release Relay   80

Section 5 AC Reduced Voltage Starters

Unit 23   Primary Resistor-Type Starters   82
Unit 24   Autotransformer Starters   91
Unit 25   Part Winding Motor Starters   94
Unit 26   Automatic Starters for Star-Delta Motors   97

Section 6 Three-Phase, Multispeed Controllers

Unit 27   Controllers for Two-Speed, Two-Winding (Separate Winding) Motors   101
Unit 28   Two-Speed, One-Winding (Consequent Pole) Motor Controller   105
Unit 29   Four-Speed, Two-Winding (Consequent Pole) Motor Controller   111

Section 7 Wound Rotor (Slip Ring) Motor Controllers

Unit 30   Manual Speed Control 116
Unit 31   Pushbutton Speed Selection 119
Unit 32   Automatic Acceleration 121
Unit 33   Automatic Speed Control 125

Section 8 Synchronous Motor Controls

Unit 34   Synchronous Motor Operation 127
Unit 35   Pushbutton Synchronizing 131
Unit 36   Timed Semiautomatic Synchronizing 133
Unit 37   Synchronous Motor Starter with Polarized Field Frequency Relay 135

Section 9 Direct-Current Controllers

Unit 38   Control Relays 141
Unit 39   Across-the-Line Starting 143
Unit 40   Use of Series Starting Resistance 145
Unit 41   Manual Faceplate Starters 147
Unit 42   Counter Emf Controller 150
Unit 43   Magnetic Time Limit Controller 152
Unit 44   Voltage Drop Acceleration 154
Unit 45   Series Relay Acceleration 156
Unit 46   Series Lockout Relay Acceleration 158
Unit 47   Dashpot Motor Control 160
Unit 48   Pilot Motor-Driven Timer Controller 164
Unit 49   Capacitor Timing Starter 166

Section 10 Methods of Deceleration

Unit 50   Jogging (Inching) Control Circuits 168
Unit 51   Plugging 172
Unit 52   Electric Brakes 179
Unit 53   Dynamic Braking 183
Unit 54   Electric Braking 188

Section 11 Motor Drives

Unit 55   Direct Drives and Pulley Drives 192
Unit 56   Gear Motors 197
Unit 57   Variable Frequency Drives 200
Unit 58   Magnetic Clutch and Magnetic Drive 202
Unit 59   Dc Variable Speed Control — Motor Drives 206

Appendix   Glossary of Terms 210
Appendix   General Information 216
Appendix   Aids to Servicing Electric Control Equipment 218
Index 223

electric motor control

BASIC
MOTOR CONTROL I
MIKE WHITE

NO SPEED
VARY
SQUIRR

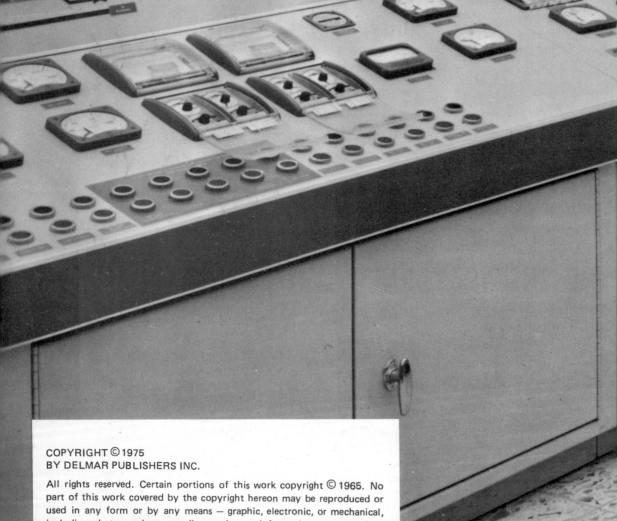

electric motor control

LIBRARY OF CONGRESS CATALOG CARD NUMBER: 73-13484

Printed in the United States of America
Published simultaneously in Canada by
Delmar Publishers, A Division of
Van Nostrand Reinhold, Ltd.

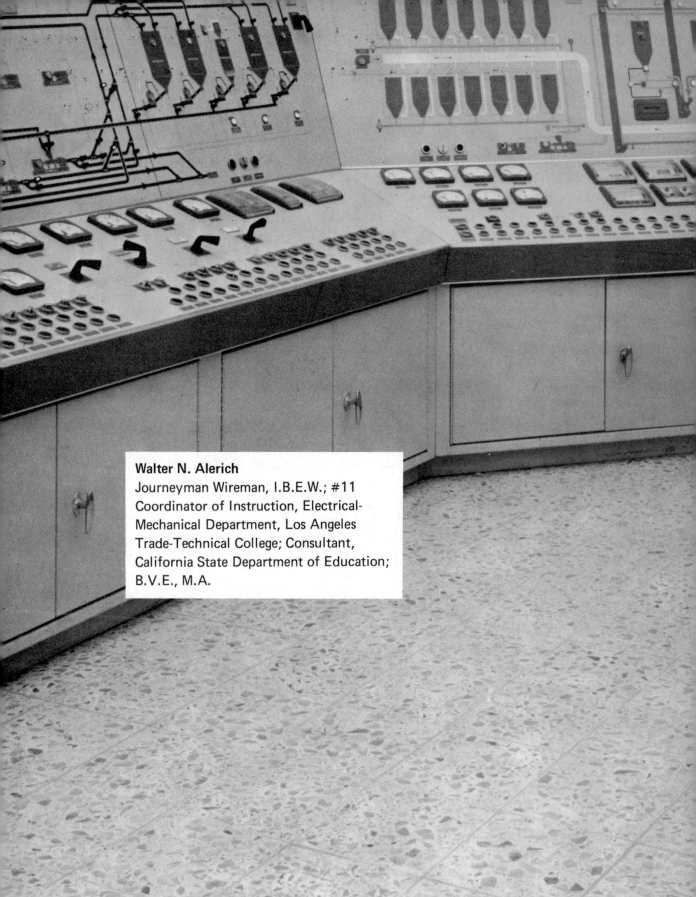

Walter N. Alerich
Journeyman Wireman, I.B.E.W.; #11
Coordinator of Instruction, Electrical-
Mechanical Department, Los Angeles
Trade-Technical College; Consultant,
California State Department of Education;
B.V.E., M.A.

DELMAR PUBLISHERS INC. • ALBANY, NEW YORK 12205

Preface

Electric motors provide one of the principal sources of power for driving machine tools and other industrial equipment. The motor is so closely related with the driving of machine elements that in most cases the motor is included as an integral part of the design of the machine.

This relationship of motor and machine through direct drive has focused attention on the design, construction, installation, and maintenance of equipment to control the motor.

The phrase *motor control* refers to the functions available from a motor controller as applied to a motor, such as speed control, reversal, acceleration, deceleration, starting, and stopping.

While various electrical texts and handbooks and manufacturers' manuals describe different types of controllers, it was often difficult to find a single source for this information. This text, however, furnishes students, apprentices, journeyman electricians, technicians, engineers, electrical contractors, instructors and others with a convenient source of related technical and practical information which is basic to a complete understanding of the theory, operation, installation, and maintenance of electric motor controllers.

Electric Motor Control is especially helpful to construction electricians and apprentices for installing and operating control systems, to plant maintenance personnel for maintaining the operation of production machines, and to draftsmen and designers of electromagnetic control systems. This text is being used in apprentice training programs, vocational schools, journeyman training, and two-year colleges in classes ranging from 54 to 108 hours.

Industrial experience requiring a working knowledge of direct- and alternating-current fundamentals and direct- and alternating-current motor theory and operation, or equivalent introductory courses, is a prerequisite to the study of this text.

The units of this text cover the common types and applications of electric motor starters and controllers, and are written in easy-to-understand language. Each unit of the text covers a concise topic. Expected student behavior is outlined in the objectives at the beginning of each unit and tested through the Study/Discussion Questions at the end of each unit. The student is encouraged to refer to the Appendix for terms and servicing aids.

The author, Walter N. Alerich, B.V.E., M.A., is the Coordinator of Instruction, Electrical-Mechanical Department, Los Angeles Trade-Technical College. He has spent many years teaching motor control classes. As a journeyman wireman, he is very knowledgeable in the practical applications of motor control. In addition, as a qualified vocational teacher, supervisor and administrator, he is close to the problem of effective instruction in this field.

The author expresses his appreciation to the officers and members of I.B.E.W. No. 11, Los Angeles, who were most helpful in constructive guidance, as were the members of the California State Curriculum Committee for Electrical Trades, which Mr. Alerich serves as Consultant.

A *Laboratory Manual – Electric Motor Control* by Walter N. Alerich is available. With the use of this Manual, which is correlated to the text, an instructor can provide students with up to 54 additional hours of related motor control application. Students will actually do what they have learned from the text. Instructions for building the portable control panels are contained in the Laboratory Manual. If desired, control panels may be purchased from either the Brodhead-Garrett Company of Cleveland, Ohio or Electronic Training Materials, Inc., P.O. Box 30172, Birmingham, Alabama, 35222.

Section 1 Introduction

Unit 1 General Principles of Electric Motor Control

OBJECTIVES

After studying this unit, the student will be able to

- State the purpose and general principles of electric motor control.
- State the difference between manual and remote control.
- List the conditions of starting and stopping, speed control, and protection of electric motors.
- Explain the difference between compensating and definite time delay action.

There are certain conditions that must be considered when selecting, designing, installing, or maintaining electric motor control equipment.

Motor control was a simple problem when motors were used to drive a common line shaft to which several machines were connected. In these cases, the motor had to be started and stopped only a few times a day. However, with individual drive, the motor is now almost an integral part of the machine, and it is necessary to design the motor controller to fit the needs of the machine to which it is connected. Large installations and the problems of starting motors in these situations may be seen in figures 1-1 and 1-2.

Fig. 1-1 Five 2000-h.p., 1800-rpm induction motors driving water pumps for a Texas oil/water flood operation. Pumps are used to force water into the ground and "float" oil upward. (Courtesy Electric Machinery Mfg. Co.)

Motor control is a broad term that means anything from a simple toggle switch to a complex system with components such as relays, timers, and switches. The common function of all controls, however, is to control the operation of an electric motor. As a result, when motor control equipment is selected and installed, many factors must be considered to insure that the motor control will function properly for the motor and the machine for which it is selected.

PURPOSE OF CONTROLLER

Some of the complicated and precise automatic applications of electrical control are illustrated in figures 1-3 and 1-6, page 5. The factors to be considered when selecting and installing motor control components for use with particular machines or systems are described in the following paragraphs.

Fig. 1-2 Horizontal 4000-h.p. synchronous motor driving large centrifugal air compressor. (Courtesy Electric Machinery Mfg. Co.)

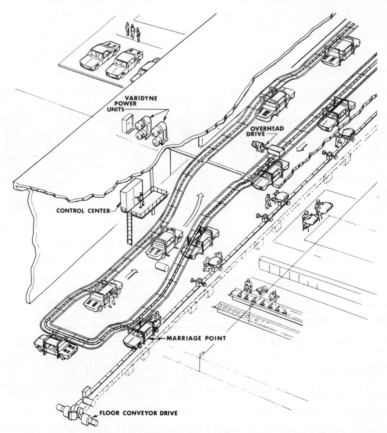

Fig. 1-3 Synchronizing two automobile assembly systems. (Courtesy U.S. Electrical Motors)

Starting

The motor may be started by connecting it directly across the source of voltage. However, the driven machine may be damaged if it is started with a sudden turning effort. Slow and gradual starting may be required, not only to protect the machine, but also to insure that the line current inrush on starting is not too great. The frequency of starting a motor is another factor affecting the controller. A motor starter is shown in figure 1-4.

Stopping

Controllers allow motors to coast to a standstill. They also impose braking action when the machine must stop quickly. Quick stopping is a vital function of the controller for emergency stops. Controllers assist the stopping action by retarding the centrifugal motion of machines and the lowering operations of crane hoists.

Fig. 1-4 Combination fused disconnect switch and motor starter (Courtesy Square D Co.)

Reversing

Controllers are required to change the direction of rotation of machines automatically or at the command of an operator at a control station. The reversing action of a controller is a continual process in many industrial applications.

Running

The maintenance of desired operational speeds and characteristics is a prime purpose and function of controllers. They protect motors, operators, machines, and materials while running.

Speed Control

Some controllers can maintain very precise speeds for industrial processes. Other controllers can change the speeds of motors either in steps or gradually through a continuous range of speeds.

Safety of Operator

Many mechanical safeguards have been replaced by electrical means of protection. Electrical control pilot devices in controllers provide a direct means of protecting machine operators from unsafe conditions.

Protection from Damage

Part of the operation of an automatic machine is to protect the machine itself and the manufactured or processed materials it handles from damage. For example, a particular machine control function may be to prevent conveyor pile-ups. A machine control can reverse, stop, slow, or do whatever is necessary to protect the machine.

Maintenance of Starting Requirements

Once properly installed and adjusted, motor starters will provide reliable maintenance of starting time, voltages, current, and torques for the benefit of the driven machine and the power system. The National Electrical Code, supplemented by local codes, governs the selection of the proper sizes of starting fuses, circuit breakers, and disconnect switches for specific system requirements.

MANUAL CONTROL

The motor may be controlled manually using any one of the following devices.

Fig. 1-5 Drum controller with cover (left) and with cover removed (right). (Courtesy Cutler-Hammer Inc.).

Toggle Switch

Many small motors are started with toggle switches. This means the motor is started directly without the use of magnetic switches or auxiliary equipment. Motors started with toggle switches are protected by the branch circuit fuse or circuit breaker. These motors generally drive fans, blowers, or other light loads. 30 Hp SMALLEST

Safety Switch - MANUAL OPERATION

It is permissible in some cases to start a motor directly across the full line voltage if an externally-operated safety switch is used. The motor receives starting and running protection from dual-element, time-delay fuses. The use of a safety switch for starting means manual operation. A safety switch, therefore, has the same limitations common to most manual starters.

Drum Controller

Drum controllers are rotary, manual switching devices which are often used to reverse motors and to control the speed of ac and dc motors. These controllers may be used without other control components in small motors, generally those with fractional horsepower ratings. 1 Hp Drum controllers are used with magnetic starters in larger motors. A drum controller is shown in figure 1-5.

Faceplate Control

Faceplate controllers have been in use for many years to start dc motors. They are also used for ac induction motor speed control. The faceplate control has multiple switching contacts mounted near a selector arm on the front of an insulated plate. Additional resistors are mounted on the rear to form a complete unit. The use of faceplate starters offers advantages and features not obtained with other manual controllers.

Fig. 1-6 Typical cement mill computer console. (Courtesy Electric Machinery Mfg. Co.)

REMOTE AND AUTOMATIC CONTROL

The motor may be controlled by remote control using pushbuttons. When pushbutton remote control is used or when automatic devices do not have the electrical capacity to carry the motor starting and running currents, magnetic switches must be included. If the motor is to be automatically controlled, the following devices may be used.

Float Switch

The raising or lowering of a float which is mechanically attached to electrical contacts may start motor-driven pumps to empty or fill tanks. Float switches are also used to open or close piping valves to control fluids.

Pressure Switch — Air Compressors

Pressure switches are used to control the pressure of liquids and gases (including air) within a desired range. Air compressors, for example, are started directly or indirectly on a call for more air by a pressure switch.

Time Clock

Time clocks can be used when a definite "on and off" period is required and adjustments are not necessary for long periods of time. A typical requirement is a motor that must start every morning at the same time and shut off every night at the same time.

Thermostat

In addition to pilot devices sensitive to liquid levels, gas pressures, and time of day, thermostats sensitive to temperature changes are widely used. Thermostats indirectly control large motors in air conditioning systems and in many industrial applications to maintain the desired temperature range of air, gases, liquids, or solids. There are many types of thermostats and temperature-actuated switches.

Limit Switch ⟶ USED in control Ckts of MAG. STARTERS

START - STOP OR Reverse of Motors

Limit switches are used most frequently as overtravel stops for machines, equipment, and products in process. These devices are used in the control circuits of magnetic starters to govern the starting, stopping, or reversal of electric motors.

Electrical or Mechanical Interlock from Other Machines

Many of the electrical control devices described in this unit can be connected in an interlocking system so that the final operation of one or more motors depends upon the electrical position of each individual control device. For example, a float switch may call for more liquid but will not be satisfied until the prior approval of a pressure switch or time clock is obtained. To design, install, and maintain electrical controls in any electrical or mechanical interlocking system, the electrical technician must understand the total operational system and the function of the individual components. It is possible with practice to transfer knowledge of circuits and descriptions to obtain an understanding of additional similar controls. It is impossible in instructional materials to show all possible combinations of an interlocking control system. However, by understanding the basic functions of control components and their basic circuitry, and by taking the time to trace and draw circuit diagrams, difficult interlocking control systems can be made easier to understand.

STARTING AND STOPPING

In starting and stopping a motor and its associated machinery, there are a number of conditions which may affect the motor.

Frequency of Starting and Stopping

The starting duty cycle of a controller is an important factor in determining how satisfactory the controller is for a particular application. Magnetic switches, such as motor starters, relays, and contactors, actually beat themselves apart from repeated opening and closing. An experienced electrician soon learns to look for this type of component failure when troubleshooting inoperative control panels. Therefore, when the frequency of starting the controller is great, the use of heavy duty controllers and accessories should be considered. For standard duty controllers, more frequent inspection and maintenance schedules should be followed.

Light or Heavy Duty Starting

Some motors can be started with no loads and others can be started with heavy loads. When motors are started, large feeder line disturbances may be created which can affect the electrical distribution system of the entire industrial plant. The disturbances may even affect the power company's system. As a result, the power companies and electrical inspection agencies place certain limitations on motor starting.

Fast or Slow Start

To obtain the maximum twisting effort of the rotor of an ac motor, the best starting condition is to apply full voltage to the motor terminals. The driven machinery, however,

Fig. 1-7   Two 1,500-h.p. vertical induction motors driving pumps. (Courtesy Electric Machinery Mfg. Co.)

Fig. 1-8   Typical 30-inch brake. (Courtesy Cutler-Hammer Inc.)

may be damaged by the sudden surge of motion. To prevent this type of damage to machines, equipment, and processed materials, some controllers are designed to start motors slowly and then increase the speed gradually in definite steps.

Smooth Starting

Although reduced electrical and mechanical surges can be obtained with a step-by-step motor starting method, very smooth and gradual starting will require different controlling methods.

Manual or Automatic Starting and Stopping

While the manual starting and stopping of machines by an operator is still a common practice, many machines and industrial processes are started and restarted automatically. These automatic devices result in tremendous savings of man-hours and materials. Automatic stopping devices are used in motor control systems for the same reasons. Automatic stopping devices greatly reduce the hazards of operating some types of machinery, both for the operator and the materials being processed. An electrically-operated mechanical brake is shown in figure 1-8.

Quick Stop or Slow Stop

Many motors are allowed to coast to a standstill. However, manufacturing requirements and safety considerations often make it necessary to bring machines to as rapid a stop as possible. Automatic controls can retard and brake the speed of a motor and also apply a torque in the opposite direction of rotation to bring about a rapid stop. The control of deceleration is one of the important functions of a motor control.

Accurate Stops

An elevator must stop at precisely the right location so that it is aligned with the floor level. Such accurate stops are possible with the use of automatic devices interlocked with control systems.

Frequency of Reversals Required

Frequent reversals of the direction of rotation of the motor impose large demands on the controller and the electrical distribution system. Special motors and special starting and running protective devices may be required to meet the conditions of frequent reversals.

SPEED CONTROL OF MOTORS

The speed control of a motor is concerned not only with starting the motor but also with controlling the motor speed while it is running. There are a number of conditions to be considered for speed control.

Constant Speed

Constant speed motors are used on water pumps. The maintenance of constant speed is essential for motor generator sets under all load conditions. Constant speed motors with speed ratings as low as 80 rpm and horsepower ratings ranging up to 5000 h.p. are used in direct drive units.

Varying Speed

A varying speed is usually preferred for cranes and hoists. In this type of application, the motor speed slows as the load increases and speeds up as the load decreases.

Fig. 1-9 Multiple synchronous motors of 3,000 h.p. and 225 rpm driving water pumps.
(Courtesy Electric Machinery Mfg. Co.)

Adjustable Speed

With adjustable speed controls, an operator can adjust the speed of a motor gradually over a wide range while the motor is running. The speed may be preset, but once it is adjusted it remains essentially constant at any load within the rating of the motor.

Multispeed

For multispeed motors, such as the type used on turret lathes, the speed can be set at two or more definite rates. Once the motor is set at a definite speed, the speed will remain practically constant regardless of load changes.

PROTECTIVE FEATURES

The particular application of each motor and control installation must be considered to determine what protective features are required.

Fig. 1-10 Large traveling overhead crane. (Courtesy Square D Co.)

Overload Protection

Running protection and overload protection refer to the same thing. A controller with overload protection will protect a motor while allowing the motor to achieve its maximum available power under a range of overload and temperature conditions. An overload may be caused by an overload on driven machinery, by a low line voltage, or by an open line in a polyphase system resulting in single-phase operation.

Open Field Protection

Dc shunt and compound-wound motors can be protected against the loss of field excitation by field loss relays. Other protective arrangements are used with starting equipment for dc and ac synchronous motors. Some sizes of dc motors may race dangerously with the loss of field excitation, while other motors may not race due to friction and the fact that they are small. *D.C. motors RUN-AWAY with loss of Field excitation*

Open-Phase Protection

Phase failure in a three-phase circuit may be caused by a blown fuse, an open connection, or a broken line. If phase failure occurs when the motor is at a standstill, the stator currents will rise to a very high value and will remain there, but the motor will remain stationary. Since the windings are not properly ventilated while the motor is stationary, the heating produced by the high currents may damage the windings. Dangerous conditions also are possible while the motor is running.

Reversed Phase Protection

If two phases of the supply of a three-phase induction motor are interchanged (phase reversal), the motor will reverse its direction of rotation. In elevator operation and industrial applications, this reversal can result in serious damage. Phase failure and phase reversal relays are used to protect motors, machines, and personnel from the hazards of open-phase or reversed-phase conditions.

Overtravel Protection

Control devices are used in magnetic starter circuits to govern the starting, stopping, and reversal of electric motors. These devices can be used to control regular operation or they can be used as emergency switches to prevent the improper functioning of machinery.

Overspeed Protection

Excessive motor speeds can damage a driven machine, materials in the industrial process, or the motor. Overspeed protection is provided in control equipment for applications such as paper and printing plants, steel mills, processing plants, and the textile industry.

Reversed Current Protection

The accidental reversal of currents in direct-current controllers can have serious effects. Direct-current controllers used with three-phase alternating-current systems which experience phase failures and phase reversals are also subject to damage. Reverse current protection is an important provision for battery charging equipment.

Mechanical Protection

For some applications an enclosure can increase the life span and contribute to the trouble-free operation of a motor and controller. Enclosures with particular ratings such as general purpose, watertight, dustproof, explosion proof, and corrosion resistant are used for specific applications, figure 1-11. All enclosures must meet the requirements of national and local electrical codes and building codes.

Short Circuit Protection

For large motors with greater than fractional horsepower ratings, short circuit protection generally is installed in the same enclosure as the motor-disconnecting means. Overcurrent devices such as fuses and circuit breakers are used to protect the motor branch circuit conductors, the motor control apparatus, and the motor itself against a sustained overcurrent due to short circuits and grounds, and prolonged and excessive starting currents.

Fig. 1-11   Spin-on, explosion-proof enclosure for a combination disconnect switch and magnetic motor starter. (Courtesy Square D Co.)

CLASSIFICATION OF AUTOMATIC CONTROL SYSTEMS

The numerous types of automatic starting and control systems are grouped into the following classifications.

Current Limiting Acceleration (also called Compensating Time)

This refers to the amount of current or voltage drop required to open or close magnetic switches. The rise and fall of the current or voltage determines a timing period which is used mainly for dc motor control. Examples of types of current limiting acceleration are:

- Counter emf or voltage drop acceleration
- Lockout contactor or series relay acceleration

Time Delay Acceleration

For this classification, *definite time* relays are used to obtain a preset timing period. Once the period is preset, it does not vary regardless of current or voltage changes occurring during motor acceleration. The following timers and timing systems are used for motor acceleration; some are also used in interlocking circuits for automatic control systems.

- Individual dashpot relays
- Multicircuit dashpot relays
- Pneumatic timing
- Inductive time limit acceleration
- Motor-driven timers
- Capacitor timing

STUDY/DISCUSSION QUESTIONS

1. What is a controller and what is its function? (Use the glossary in the Appendix and the information from this unit to answer this question.)

2. What is remote control?

3. To what does current limiting, or compensating time, acceleration refer?

4. List some devices that are used to control a motor automatically. Briefly describe the purpose of each device.

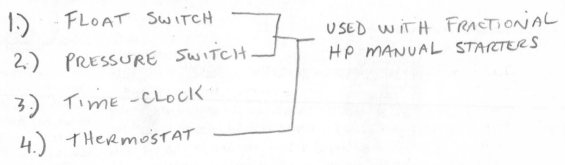

1.) FLOAT SWITCH ⎤
2.) PRESSURE SWITCH ⎦⎤ USED WITH FRACTIONAL HP MANUAL STARTERS
3.) TIME-CLOCK
4.) THERMOSTAT

Unit 2 Fractional Horsepower Manual Motor Starters

OBJECTIVES

After studying this unit, this student will be able to

- Match simple schematic diagrams with the appropriate manual motor starters.
- Connect manual fractional horsepower motor starters for automatic and manual operation.
- Explain the principles of operation of manual motor starters.
- Read and draw simple schematic diagrams.

One of the simplest types of motor starters is an on-off, snap action switch operated by a toggle lever mounted on the front of the starter. The motor is connected directly across the line voltage on start. This situation usually is not objectionable with motors rated at one horsepower (h.p.) or less. Since a motor may draw up to a 600-percent current surge on starting, larger motors should not be connected directly across the line on startup. Such a connection would result in large line surges that may disrupt power services or cause voltage fluctuations which impede the normal operation of other equipment.

Fractional horsepower manual motor starters are used whenever it is desired to provide overload protection for a motor as well as off and on control of small alternating-current single-phase or direct-current motors. Electrical codes require that fractional horsepower motors be provided with overload protection whenever they are started automatically or by remote control.

Since manual starters are manually-operated mechanical devices, the contacts remain closed and the lever stays in the ON position in the event of a power failure. As a result, the motor automatically restarts when the power returns. Therefore, low-voltage protection and low-voltage release (see Glossary in the Appendix) are not possible with manually-operated starters. This automatic restart action is an advantage when the starter is used with motors that run continuously, such as those used on fans, blowers, pumps, and oil burners.

Fig. 2-1 Open-type starter with overload heater.
(Courtesy Square D Co.)

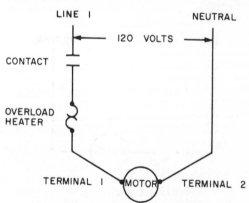

Fig. 2-2 Diagram of single-pole manual starter shown in figure 2-1.

The compact construction of the manual starter means that it requires little mounting space and can be installed on the driven machinery and in various other places where the available space is limited. The unenclosed, or open starter, can be mounted in a standard switch or conduit box installed in a wall and can be covered with a standard flush, single-gang switch plate. The ON and OFF positions are clearly marked on the operating lever, which is very similar to a standard lighting toggle switch lever, figure 2-1.

Fractional horsepower manual starters have thermal overload protection, figure 2-2. When an overload occurs, the starter handle automatically moves to the center position to indicate that the contacts have opened and the motor is no longer operating. The starter contacts cannot be reclosed until the overload relay is reset manually. The relay is reset by moving the handle to the full OFF position after allowing about two minutes for the relay to cool.

Fractional horsepower manual starters are provided in several different types of enclosures as well as the open type. Enclosures can be obtained to shield the live starter circuit components from accidental contact, for mounting in machine cavities, to protect the starter from dust and moisture, or to prevent the possibility of an explosion when the starter is used in hazardous locations.

AUTOMATIC OPERATION

Common applications of manual starters provide control of small machine tools, fans, pumps, oil burners, blowers, and unit heaters. Almost any small motor should be controlled with a starter of this type. However, the contact capacity of the starter must be sufficient to make and break the full motor current. Automatic control devices such as pressure switches, float switches, or thermostats may be used with fractional horsepower manual starters.

The schematic diagram shown in figure 2-3 illustrates a fractional horsepower motor controlled automatically by a float switch that is connected in the small motor circuit as long as the manual starter contact is closed.

In figure 2-4, the selector switch must be turned to the automatic position if the float switch is to take over an automatic operation, such as sump pumping. A filled sump raises the float, closes the normally open electrical contact, and starts the motor. When the motor pumps the sump or tank empty, the float lowers and breaks electrical contact with the motor, thus stopping the motor. This cycle of events will repeat when the sump fills again.

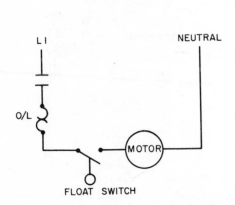

Fig. 2-3 Automatic control with manual starter for fractional h.p. motor without selector switch.

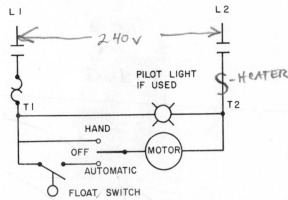

Fig. 2-4 Manual starter with automatic control for fractional h.p. motor using selector switch.

Note that a double-pole starter is used in figure 2-4. This type of starter is required when both lines to the motor must be broken. The double-pole starter is recommended for severe duty applications because of its higher interrupting capacity and longer contact life.

THERMAL OVERLOAD PROTECTION

Thermal overload units are widely used on fractional horsepower manual starters for electrical protection of motors from sustained overcurrents resulting from overloading of the driven machine or excessively low line voltage. Heating elements which are closely calibrated to the full load current of the motor cause alloy elements to melt when there is a motor overload due to excess cur-

Fig. 2-5 Fractional h.p. manual starter with selector switch, mounted in general-purpose enclosure.

rent in the circuit   When the element melts, a spring-loaded ratchet is rotated and trips open a contact which then stops the motor.

Normal motor starting currents and momentary overloads will not cause thermal relays to trip because of their inverse-time characteristics (see Glossary in the Appendix). However, continuous overcurrent through the heater unit raises the temperature of the alloy elements. When the melting point is reached, the ratchet is released and the switch mechanism is tripped to open the line or lines to the motor. The switch mechanism is trip-free, which means that it is impossible to hold the contacts closed against an overload.

Only one relay is required in either the single-pole or double-pole motor starter, since the starter is intended for use on dc or single-phase ac service. These relays offer protection against continued operation when the line current is excessively high. Relays with meltable alloy elements are nontemperable and give reliable overload protection. Repeated tripping does not cause deterioration, nor does it affect the accuracy of the trip point.

Many types of relay units are available so that the proper one can be selected on the basis of the actual full-load current rating of the motor. The relay units are interchangeable

Fig. 2-6 Line voltage manual starter with general-purpose enclosure (cover removed). (Courtesy Square D Co.)

Fig. 2-7 Line voltage manual starter (Courtesy Square D Co.)

and are accessible from the front of the starter. Since the motor current is connected in series with the heater coil, the motor will not operate unless the relay unit has the heater installed. Overload units may be changed without disconnecting the wires from the switch or removing the switch from the enclosure. However, the switch should be turned off for safety reasons.

MANUAL PUSHBUTTON LINE VOLTAGE STARTERS

Generally, manual pushbutton starters may be used to control single-phase motors rated up to 5 h.p., polyphase motors rated up to 10 h.p., and direct-current motors rated up to 2 h.p. Typical manual pushbutton starters (common and new models respectively) are shown in figures 2-6 and 2-7.

STUDY/DISCUSSION QUESTIONS

1. If the contacts on a manual starter cannot be closed immediately after a motor overload has tripped them open, what is the probable reason?

2. If the handle of an installed motor starter is in the center position, what condition does this indicate?

3. How may a manual starter be installed or used for an automatic operation?

4. What does trip-free mean?

5. If overload heating elements are not installed in the starter, what is the result?

Unit 3 Magnetic Line Voltage Starters

OBJECTIVES

After studying this unit, the student will be able to

- Identify common magnetic motor starters and overload relays.
- Describe the construction and operating principles of magnetic switches.
- Describe the operating principle of a solenoid.
- Troubleshoot magnetic switches.
- Select starter protective enclosures for particular applications.

Magnetic control means the use of electromagnetic energy to close switches. Line voltage magnetic starters are electromechanical devices that provide a safe, convenient, and economic means of starting and stopping motors. In addition, these devices can be controlled remotely. They are used where a full-voltage starting torque (see Glossary in the Appendix) may be applied safely to the driven machinery and where the current inrush resulting from across-the-line starting is not objectionable. Control for these starters usually is provided by pilot devices such as pushbuttons, float switches, or timing relays.

Magnetic starters are available in many sizes as shown in Table 3-1. Each size has been assigned a horsepower rating that applies when the motor used with the starter is used for normal starting duty. All starter ratings comply with the National Electrical Manufacturers Association Standards. The capacity of a starter is determined by the size of its contacts and the wire cross-sectional area. The size of the contacts is reduced when the voltage is doubled because the current is halved for the same power.

Three-pole starters are used with motors operating on three-phase, three-wire ac systems. Two-pole starters are used for single-phase motor starting. The number of poles refers to the power contacts, or the motor load contacts, and does not include control contacts for control circuit wiring.

The simple up-and-down motion of a solenoid-operated, three-pole magnetic switch is shown in figure 3-2. The figure does not show motor overload relays. Double break contacts are used to cut the voltage in half on

Table 3-1 Controller Sizes
(NEMA Designations)

Starter Size	Motor Maximum H.P., Three-Phase	Motor Voltage
00	1/3	
0	1 1/2	
1	3	110
2	7 1/2	
3	15	
4	25	
00	1 1/2	
0	2	
1	5	
1 3/4	10	
2	15	208-230
3	30	
4	50	
5	100	
6	200	
7	300	
8	450	
00	2	
0	3	
1	7 1/2	
2	25	
3	50	
4	100	440-550
5	200	
6	400	
7	600	
8	900	

each contact, thus providing high arc-rupturing capacity and longer contact life.

OVERLOAD PROTECTION

Electric motor overload protection is necessary to prevent burnout and to insure the maximum operating life of the motor. Electric motors, if permitted, will operate at an output of more than their rated capacity. Motor overload may be caused by an overload on the driven machinery, a low line voltage, or an open line in a polyphase system resulting in

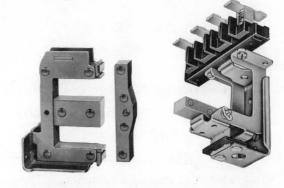

Fig. 3-1 Magnet structure (left) and movable contacts and armature guide assembly (right) of a four-pole magnetic switch. (Courtesy Square D Co.)

single-phase operation. Under any condition of overload, a motor draws excessive current that causes overheating. Since the insulation of motor windings deteriorates when subjected to overheating, there are established limits on motor operating temperatures.

To provide *overload* or *running protection* to keep a motor from overheating, overload relays are used on starters to limit to a predetermined value the amount of current drawn. Local electrical codes determine the size of protective overload relays.

These relays have current-sensitive thermal or magnetic elements which are connected either directly in the motor lines or indirectly in the motor lines through current transformers. When excessive current is drawn, the relay deenergizes the starter and stops the motor.

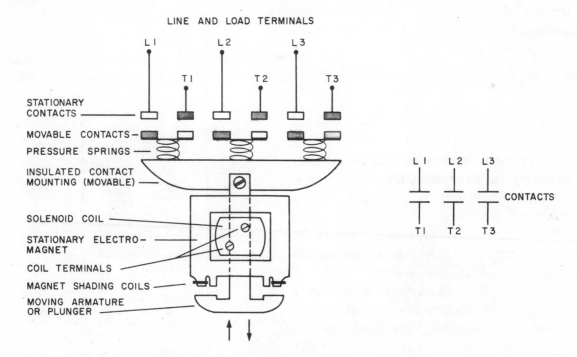

Fig. 3-2 Three-pole, solenoid-operated magnetic switch.

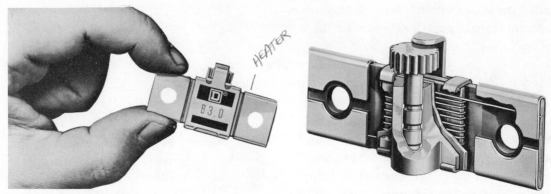

Fig. 3-3 Melting alloy type overload heater. Cutaway view (right) shows construction of heater. (Courtesy Square D Co.)

Melting Alloy Thermal Units – MORE ACCURATE RELAY – NOT POSSIBLE TO CLOSE MANUALLY.

The melting alloy assembly consisting of a heater element and solder pot is shown in figure 3-3. Excessive motor current passes through the heater element and melts an alloy solder pot. Since the ratchet wheel is then free to turn in the molten pool, it trips the starter control circuit to stop the motor. A cooling-off period is required to allow the solder pot to become solid again before the overload relay can be reset and motor service restored.

Melting alloy thermal units are interchangeable. They have a one-piece construction which insures a constant relationship between the heater element and the solder pot. As a result, this unit can be factory calibrated, making it virtually tamper-proof in the field. These important features are not possible with any other type of overload relay construction. A wide selection of interchangeable thermal units is available to give exact overload protection to motors of any full-load current ratings.

Bimetallic Overload Relays

These devices are designed specifically for two general types of application. First, the automatic reset feature means that the devices can be mounted in locations which are not

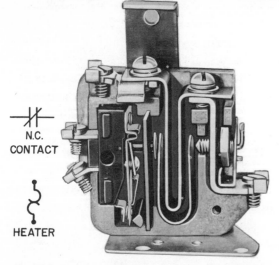

Fig. 3-4 Cutaway view showing construction of bimetallic overload relay (Courtesy Square D Co.)

Fig. 3-5 Fluid dashpot timing relay (Courtesy Square D Co.)

easily accessible for manual operation. Second, bimetallic relays can be adjusted to trip within a range of 85 to 115 percent of the nominal trip rating of the heater unit. This feature is useful when the recommended heater size may result in unnecessary tripping, while the next larger size will not give adequate protection. Ambient temperatures (see Glossary in the Appendix) affect thermally-operated overload relays.

The tripping of the control circuit in the bimetallic relay results from the difference of expansion of two dissimilar metals fused together. Movement occurs when one of the metals expands more than the other when subjected to heat. A U-shaped bimetallic strip, figure 3-4, is used to calibrate this type of relay. The U-shaped strip and a heater element inserted in the center of the U compensate for uneven heating due to variations in the mounting location of the heater element. Since a motor starter is installed in series with the load, the starter must have the heating element (either bimetallic or solder pot) installed in the overload relay before a motor will start.

Magnetic Overload Relays

The magnetic overload relay is connected in series directly with the motor or is indirectly connected (in circuits with large motors) by the use of current transformers. As a result, the coil of the magnetic relay must be wound with wire large enough in size to pass the motor current.

Magnetic overload relays are used when an electrical contact must be opened or closed whenever the actuating current rises to a certain value. In some cases, the relay may also be used so that it is actuated when the current falls to a certain value. Magnetic overload relays are used to protect large motor windings against continued overcurrent. Typical applications are to stop a material conveyor when conveyors ahead become overloaded, and to limit torque reflected by the motor current.

Time Limit Overload Relays

Time delay overload relays, figure 3-5, make use of the dashpot principle. Motor current passing through the coil of the relay exerts a magnetic pull on a plunger. This pull tends to raise the plunger which is attached to a piston immersed in oil. As the current increases in the relay coil, the force of gravity is overcome and the plunger and piston move upward. During this upward movement, oil is forced through bypass holes in the piston. As a result, the operation of the contacts is delayed. A valve disc is turned to open or close bypass holes of various sizes in the piston. This action changes the rate of oil flow and thus adjusts the time delay characteristic. The rate of upward travel of the core and piston depends directly upon the degree of overload; the greater the current load, the faster the upward movement. As the rate of upward movement increases, the relay tripping time decreases.

This inverse time characteristic prevents the relay from tripping on the normal starting current or on harmless momentary overloads. In these cases, the line current resumes its normal value before the operating coil is able to lift the core and piston far enough to operate the contacts. However, if the overcurrent continues for a prolonged period, the core is pulled far enough to operate the contacts. As the line current increases, the relay tripping time decreases. Tripping current adjustment is achieved by adjusting the plunger core with respect to the overload relay coil. Quick tripping is obtained through the use of a light grade dashpot oil and adjustment of the oil bypass holes.

A valve in the piston allows almost instantaneous resetting of the circuit to restart the motor. The current then must be reduced to a very low value before the relay will reset. This action is accomplished automatically when the tripping of the relay disconnects the motor from the line. Magnetic overload relays are available with either automatic reset contacts or hand reset contacts.

Instantaneous Trip Current Relays

Instantaneous trip current relays are used to take a motor off the line as soon as a predetermined load condition is reached. For example, when a blockage of material on a woodworking machine causes a sudden high current, an instantaneous trip relay can cut off the motor quickly. After the cause of the blockage is removed, the motor can be restarted immediately because the relay resets itself as soon as the overload is removed. This type of relay also is used on conveyors to stop the motor before mechanical breakage results from a blockage.

The instantaneous trip current relay does not have the inverse time characteristic. Thus, it must not be used in ordinary applications requiring an overload relay. The instantaneous trip current relay should be considered as a special-purpose relay.

The operating mechanism of the trip relay in figure 3-6 consists of a solenoid coil through which the motor current flows. There is a movable iron core within the coil. Mounted on top of the solenoid frame is a snap-action precision switch which has connections for either a normally open or normally closed contact. The motor current exerts a magnetic pull upward on the iron core.

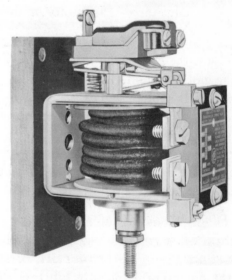

Fig. 3-6 Instantaneous trip current relay.
(Courtesy Allen-Bradley Co.)

Normally, however, the pull is not sufficient to lift the core. If an overcurrent condition causes the core to be lifted, the snap-action precision switch is operated to trip the relay.

The tripping value of the relay can be set over a wide range of current ratings by moving the plunger core up and down on the threaded stem. As a result, the position of the core in the solenoid is changed. By lowering the core, the magnetic flux is weakened and a higher current is required to lift the core and trip the relay.

Number of Overload Relays Needed to Protect a Motor

There are more three-phase motor starters with two overload relays already installed than starters with three overload relays. However, this trend will probably reverse due to a National Electrical Code requirement of three overload relays for three-phase starters on new installations.

A balanced supply voltage must be maintained for all polyphase load installations.

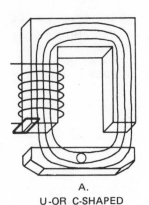
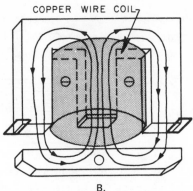
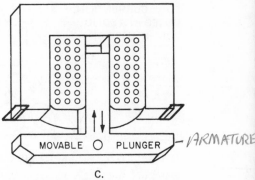

COPPER WIRE COIL

MOVABLE ○ PLUNGER — *ARMATURE*

A.
U-OR C-SHAPED
MAGNET CORE

B.
BALANCED MAGNETIC ATTRACTION
IS POSSIBLE WITH E-SHAPED IRON CORE

C.
SOLENOID DESIGN IS A VARIATION
OF E-SHAPED ELECTROMAGNET

Fig. 3-7 Variations of basic magnet core and coil configurations

A single-phase load on a three-phase circuit can produce serious unbalanced motor currents. A large three-phase motor on the same feeder with a small three-phase motor may not be protected if a single-phase condition occurs.

A loose or broken wire anywhere in the conduit system or in a motor lead can result in single-phase operation. This will show up as a sluggish, hot-running motor. Sometimes the motor will not start at all but will produce a distinct magnetic hum when it is energized. This is also a sign of a single-phase condition in a three-phase motor.

Unbalanced single-phase loads on three-phase panel boards must be avoided. Problems may occur on distribution systems where one or more large motors may feed back power to smaller motors under open-phase conditions.

STARTER ELECTROMAGNETS

Electrical control equipment makes extensive use of a device called a solenoid. This electromechanical device is used to operate motor starters, contactors, relays, and valves. By placing a coil of many turns of wire around a soft iron core, the magnetic flux set up by the energized coil tends to be concentrated; therefore, the magnetic field effect is strengthened. Since the iron core is the path of least resistance to the magnetic lines of force, magnetic attraction concentrates according to the shape of the magnet core.

There are several variations in design of the basic solenoid magnetic core and coil, figure 3-7.

As shown in the solenoid design of figure 3-7C, linkage to movable contacts is obtained through a hole in the movable plunger. The plunger is shown in the open position in the figure.

The center leg of each of the E-shaped magnet cores in figures 3-7B and C is ground shorter than the outside legs to prevent the accidental contact of the magnet and the plunger due to residual magnetism when power is

STATIONARY
MOVABLE

Fig. 3-8 Ac magnetic starter with contact arcing chamber removed. Note overload relay (lower left) and coil and magnet (lower center). (Courtesy Square D Co.)

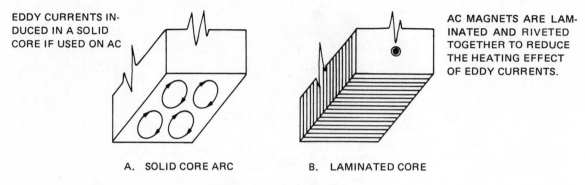

A. SOLID CORE ARC   B. LAMINATED CORE

Fig. 3-9 Types of magnet cores

disconnected. The OFF or OPEN position is obtained by deenergizing the coil and allowing the force of gravity or spring tension to release the plunger from the magnet body, thereby opening the electrical contacts. The actual contact surfaces of the plunger and core body are machine ground to insure a high degree of flatness on the contact surfaces so that operation on alternating current is quieter. Improper alignment of the contacting surfaces and foreign matter between the surfaces may cause a noisy hum.

Another source of noise is due to loose laminations. The magnet body and plunger (armature) are made up of thin sheets of metal laminated and riveted together to reduce *eddy currents*. Eddy currents are shorted currents induced in the metal by the transformer action of an ac coil. Although these currents are small, they heat up the metal, create an iron loss, and contribute to inefficiency. Laminations in early magnets were insulated from each other by a thin, nonmagnetic coating; however, it was found that the normal oxidation of the metallic laminations reduces the effects of eddy currents to a satisfactory degree, thus eliminating the need for a coating.

SHADED POLE PRINCIPLE

The shaded pole principle is used to provide a time delay in the decay of flux in dc coils, and to prevent chatter and wear in the moving parts of ac magnets.

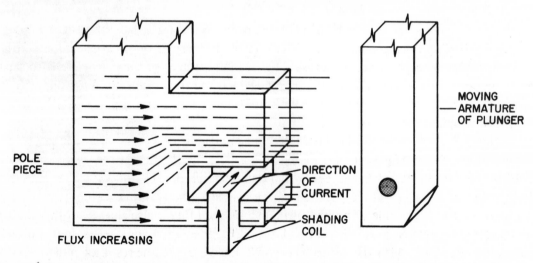

Fig. 3-10 Pole face section with shading coil; current is in the clockwise direction for increasing flux.

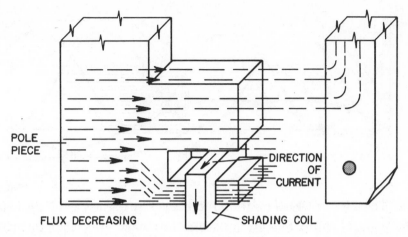

Fig. 3-11 Pole face section; current is in the counterclockwise direction for decreasing flux

Time Delay

Figure 3-10 shows a copper band or short-circuited coil of low resistance connected around a portion of a magnet pole piece. When the flux is increasing in the pole piece from left to right, the induced current in the shading coil is in a clockwise direction.

The magnetic flux produced by the shading coil opposes the direction of the flux of the main field. Therefore, with the shading coil in place, the flux density in the shaded portion of the magnet will be considerably less, and the flux density in the unshaded portion of the magnet will be more than if the shading coil were not in place.

Figure 3-11 shows the magnet pole with the flux direction still from left to right, but now the flux is decreasing in value. The current in the coil is in a counterclockwise direction. As a result, the magnetic flux produced by the coil is in the same direction as the main field flux. With the shading coil in place, the flux density in the shaded portion of the magnet will be larger and that in the unshaded portion less than if the shading coil were not used.

Thus, when the electric circuit of a coil is opened, the current decreases rapidly to zero, but the flux decreases much more slowly because of the action of the shading coil.

Use of The Shading Pole to Prevent Wear

The attraction of an electromagnet operating on alternating current is pulsating and equals zero twice during each cycle. The pull of the magnet on its armature also drops to zero twice during each cycle. As a result, the sealing surfaces of the magnet tend to separate each time the flux is zero and then contact again as the flux builds up in the opposite direction. This continual making and breaking of contact will result in a noisy starter and wear on the moving parts of the magnet.

This noise and wear can be eliminated in ac magnets by the use of shaded poles. It was shown previously in this unit that by shading a pole tip, the flux in the shaded portion lags behind the flux in the unshaded portion. Figure 3-12 shows the flux variations with time in both the shaded and unshaded portions of the magnet. The two flux waves are made as near 90 degrees apart as possible. The pull produced by each flux is also shown. If the flux waves are exactly 90 degrees apart, the pulls will be 180 degrees apart and the resultant pull will be constant. However, with the fluxes *nearly* 90 degrees apart, the resulting pull varies only a small amount from its average value and never goes through zero. The voltage induced

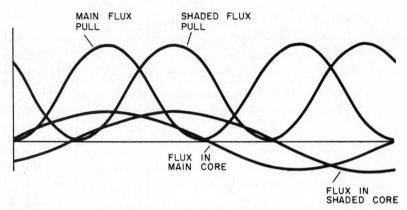

MAIN FLUX
PULL

SHADED FLUX
PULL

FLUX IN
MAIN CORE

FLUX IN
SHADED CORE

Fig. 3-12 Flux variations in both the shaded and unshaded portions of the magnet

in the shading coil causes flux to exist in the electromagnet, even when the main coil current instantaneously passes through a zero point. As a result, contact between the sealing surfaces of the magnet is not broken and chattering and wear are prevented.

AC COMBINATION STARTERS

With minor exceptions, the National Electrical Code and some local codes require that every motor have a disconnect means such as a combination starter. This device consists of an across-the-line starter and a disconnect means wired together in a common enclosure. Combination starters may have a blade-type disconnect switch, either fusible or nonfusible, or a thermal-magnetic trip circuit breaker. The starter may be controlled remotely with pushbuttons or selector switches, or these devices may be installed in the cover of the starter enclosure. The combination starter takes little mounting space and makes a compact electrical installation possible.

A combination starter provides safety for the operator because the cover of the enclosure is interlocked with the external operating handle of the disconnecting means. The door cannot be opened while the disconnecting means is closed. When the disconnect means is open, all parts of the starter are accessible; however, the hazard is reduced since the readily accessible parts of the starter are not connected to the power line. This safety feature is not available on separately enclosed starters. In addition, the starter enclosure is provided with a means for padlocking the disconnect in the OFF position.

Protective Enclosures

The selection and installation of the correct enclosure for a particular application can contribute to the length of useful service and freedom from trouble in operating electromagnetic control equipment. To shield workers and other equipment in the vicinity from accidental contact with electrically live parts some form of enclosure is necessary. This requirement is usually met by a general-purpose, sheet steel cabinet. The presence of dust, moisture, or explosive gases often makes it necessary to use a special enclosure to protect the controller from corrosion or the surrounding equipment from possible explosions. In selecting and installing control apparatus, it is necessary to consider carefully the conditions under which the apparatus must operate; there are many applications where a general-purpose sheet steel enclosure does not give sufficient protection.

Watertight and dusttight enclosures are used for the protection of control apparatus. Dirt, oil, or excessive moisture are destructive to insulation and frequently form current-carrying paths that lead to short circuits or grounded circuits.

Special enclosures for hazardous locations are used for the protection of life and property. Explosive vapors or dusts exist in some departments of many industrial plants, as well as in grain elevators, refineries, and chemical plants. The National Electrical Code and local codes describe hazardous locations. The Underwriters' Laboratories have defined the requirements for protective enclosures according to the hazardous conditions. The National Electrical Manufacturers Association (NEMA) has standardized enclosures from these requirements.

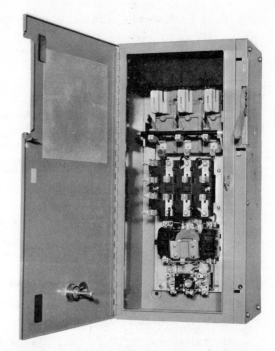

Fig. 3-13 Combination starter with fuses removed. Disconnect switch must be open before door can be opened.

General-Purpose Enclosures (NEMA 1) are constructed of sheet steel, and are designed to prevent accidental contact with live parts. Covers have latches with provisions for padlocking, figure 3-14.

Watertight Enclosures (NEMA 4) are made of cast construction or of sheet metal of suitable rigidity and are designed to pass a hose test with no leakage of water. Watertight enclosures are suitable for outdoor applications, on ship docks, in dairies, in breweries, and in other locations where the apparatus is subjected to dripping or splashing liquids, figure 3-15.

Fig. 3-14 General-Purpose enclosure (NEMA 1).

Fig. 3-15 Watertight enclosure (NEMA 4).
(All figures Courtesy Square D Co.)

Fig. 3-16 Hazardous location enclosure (NEMA 7).

EXPLOSION - PROOF

Dusttight Enclosures (NEMA 5 and 12) are constructed of sheet steel and are provided with cover gaskets to exclude dust. Dusttight enclosures are suitable for use in steel mills, coke plants, and similar locations where nonhazardous dusts are present.

Class 1, Group D Enclosures (NEMA 7) are designed for use in hazardous locations where atmospheres containing gasoline, petroleum, naphtha, alcohol, acetone, or lacquer solvent vapors are present or may be encountered. Enclosures are heavy grey iron castings, machined to provide a metal-to-metal seal, figure 3-16, page 25.

Applicable and enforced National, State, or local electrical codes and ordinances should be consulted to determine the safe way to make any installation.

STUDY/DISCUSSION QUESTIONS

1. What is a magnetic line voltage motor starter?

2. How many poles are required on motor starters for the following motors: 240-volt, single-phase induction motor; 440-volt, three-phase induction motor?

3. If a motor starter is installed according to directions, but will not start, what is a common cause for failure to start?

4. Using the time limit overload, or the dashpot overload relay, how are the following achieved: time delay characteristics; tripping current adjustments?

5. What is meant by chattering of an ac magnet?

6. What is the phase relationship between the flux in the main pole of a magnet and the flux in the shaded portion of the pole?

7. In what devices is the principle of the shaded pole used?

8. What does the electrician look for to remedy the following conditions: loud or noisy hum; chatter?

9. What type of protective enclosure is used most commonly?

10. Why is a disconnect fused switch or circuit breaker installed with a motor starter?

11. What safety feature does the type of assembly given in question 10 provide that individual starter assemblies do not?

12. List the probable causes if the armature does not release after the magnetic starter is deenergized.

13. How is the size of the overload heaters selected for a particular installation?

14. What type of motor starter enclosure is recommended for an installation requiring safe operation around an outside paint filling pump?

Section 2 Control Pilot Devices

Unit 4 Pushbutton Control

OBJECTIVES

After studying this unit, the student will be able to

- Describe the difference between normally closed and normally open pushbuttons.
- Draw the wiring diagram symbols for pushbuttons and pilot lights.

A pushbutton station is a device that provides control of a motor by pressing a button. By using more than one pushbutton station it is possible to control a motor from as many places as there are stations through the same magnetic controller.

Two sets of contacts are usually provided with pushbuttons so that when the button is pressed, one set of contacts is opened and the other set is closed. Thus, by connecting to the proper set of contacts, either a normally open or a normally closed situation is obtained. Normally open and normally closed mean that the contacts are in a "rest" position and are not subject to either mechanical or electrical external forces. (See the Glossary in the Appendix for more detailed definitions.)

Pushbutton stations are made for two types of service: standard duty stations for normal applications, and heavy duty stations when the pushbuttons are to be used frequently.

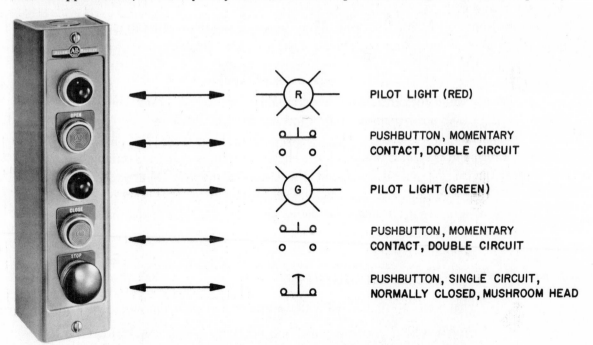

PILOT LIGHT (RED)

PUSHBUTTON, MOMENTARY
CONTACT, DOUBLE CIRCUIT

PILOT LIGHT (GREEN)

PUSHBUTTON, MOMENTARY
CONTACT, DOUBLE CIRCUIT

PUSHBUTTON, SINGLE CIRCUIT,
NORMALLY CLOSED, MUSHROOM HEAD

Fig. 4-1 Heavy duty control station, showing general-purpose enclosure
and electrical symbol for each element. (Courtesy Allen-Bradley Co.)

A. STANDARD DUTY
(Courtesy Cutler-Hammer Inc.)

B. HEAVY DUTY
(Courtesy Allen-Bradley Co.)

Fig. 4-2 Open selector switches.

Fig. 4-3 Open pushbutton unit with mush-room head. (Courtesy Allen-Bradley Co.)

The pushbutton station enclosure containing the contacts is usually made of molded plastic or sheet metal. Some contacts are made of copper, although in most pushbuttons silver to silver contacts are provided.

Since control buttons are subject to high momentary voltages caused by the inductive effect of the coils to which they are connected, good clearance between the contacts and insulation to ground is provided.

The pushbutton station may be mounted adjacent to the controller or at a distance from it. The current broken by a pushbutton is usually small. As a result, the operation of the controller is hardly affected by the length of the wires leading from the controller to a remote pushbutton station.

Pushbuttons can be used to control any or all of the many operating conditions of a motor, such as start, stop, forward, reverse, fast, and slow. Pushbuttons also may be used as remote stop buttons with manual controllers equipped with potential trip or low voltage protection.

Standard pushbutton enclosures are available for normal conditions, while special enclosures are used in situations requiring watertight, dusttight, explosion-proof, or submersible protection.

Indicating lights may be mounted in the enclosure. These lights are usually red or green and indicate when the line is energized, the motor is running, or any other designated condition. Control stations also may include selector switches that are key, coin, or hand operated. Provisions are often made for padlocking stop buttons in the open position (for safety purposes).

STUDY/DISCUSSION QUESTIONS

1. What is meant by normally open contacts and normally closed contacts?

2. Why is it that normally open and normally closed contacts cannot be closed simultaneously?

3. How are colored pilot lights indicated in wiring diagrams?

Unit 5 Relays and Contactors

OBJECTIVES

After studying this unit, the student will be able to

- Tell how magnetic relays differ from contactors and list the principal uses of each.
- Describe the operation of magnetic blowout coils and how they provide arc suppression.
- Draw the wiring diagram symbols for contactors and relays.
- Describe the operation and use of mechanically held relays.

CONTROL RELAYS

Magnetic relays are used as auxiliary devices to switch control circuits and large starter coils and to control light loads, such as small motors, electric heaters, pilot lights, and audible signal devices. Magnetic relays do not provide motor overload protection. This type of relay ordinarily is used in a two-wire control system (any electrical contact-making device with two wires). Whenever it is desired to use momentary contact pilot devices, such as push-buttons, any available normally open contact can be wired as a holding circuit in a three-wire system (see Glossary in the Appendix).

The contact arrangement and a description of the magnetic structure of relays was presented in Unit 3.

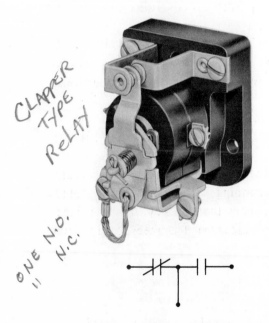

Fig. 5-1 Single-pole, double-throw, ac control relay with wiring symbol for contacts. (Courtesy Square D Co.)

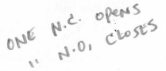

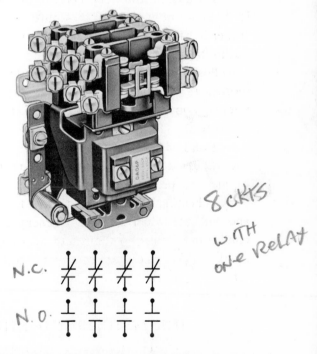

Fig. 5-2 Control relay with four normally open and four normally closed contacts. To change from the open to the closed condition, wire connections to the respective poles are changed. (Courtesy Allen-Bradley Co.)

Control relays are available in single- or double-throw arrangements with various combinations of normally open and normally closed contact circuits. Because of the variety of styles of relays available, it is possible to select the correct relay for almost any application.

Relays are used more often to open and close control circuits than to operate power circuits. Typical applications include the control of motor starter and contactor coils, the switching of solenoids, and the control of other relays. A relay is a small but vital switching component of many complex control systems. Low-voltage relay systems are used extensively in switching residential and commercial lighting circuits and individual lighting fixtures.

While control relays from various manufacturers differ in appearance and construction, they are interchangeable in control wiring systems if their specifications are matched to the requirements of the system.

CONTACTORS

Magnetic contactors are electromagnetically-operated switches that provide a safe and convenient means for connecting and interrupting branch circuits. The principal difference between a contactor and a motor starter is that the contactor does not contain overload relays. Contactors are used in combination with pilot control devices to switch lighting and heating loads and to control ac motors where overload protection is provided separately. The larger contactor sizes are used to provide remote control of relatively high-current circuits where it is too expensive to run the power leads to the remote controlling location. This flexibility is one of the main advantages of electromagnetic control over manual control. Pilot devices such as pushbuttons, float switches, pressure switches, limit switches, and thermostats are provided to operate the contactors.

Magnetic Blowout

The contactors shown in figures 5-3 and 5-4 operate on alternating current. Heavy duty contact arc-chutes are provided on these contactors. The chutes contain heavy copper coils,

Fig. 5-3 Open vertical Size 1 magnetic contactor (left), and "clapper" action Size 6 magnetic contactor (right). (Courtesy Square D Co.)

Fig. 5-4 Contactor, Size 1.
Contacts are accessible
by removing the two
front screws.
(Courtesy Square D Co.)

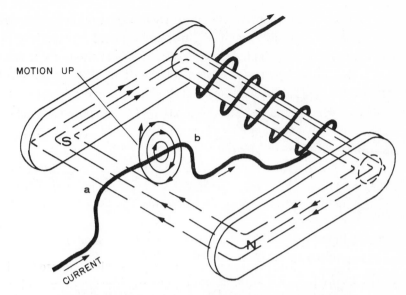

Fig. 5-5 Illustration of the magnetic blowout principle.
Straight conductor simulates arc.

called blowout coils, mounted above the contacts in series with the load to provide better arc suppression. These magnetic blowout coils help to extinguish an electric arc at contacts opening under alternating-current and direct-current loads. An arc-quenching device is used to assure longer contact life. Since the hot arc is transferred from the contact tips very rapidly, the contacts remain cool and thus last longer.

Contactor and motor starter contacts that frequently break heavy currents are subject to a destructive burning effect if the arc is not quickly extinguished. The arc formed when the contacts open can be lengthened and thus extinguished by motor action if it is in a magnetic field. This magnetic field is provided by the magnetic blowout coil. Since the coil of the magnet is usually in series with the line, the field strength and extinguishing action are in proportion to the size of the arc.

Figure 5-5 is a sketch of a blowout magnet with a straight conductor (ab) located in the field and in series with the magnet. This figure can represent either dc polarity or instantaneous ac. With ac current, the blowout coil magnetic field and the conductor (arc) magnetic field will reverse simultaneously. According to Fleming's left-hand rule, motor action will tend to force the conductor in an upward direction. The application of the right-hand rule for a single conductor shows that the magnetic field around the conductor aids the main field on

the bottom of the conductor and opposes it on the top, thus producing an upward force on the conductor.

Figure 5-6 shows a section of figure 5-5 with the wire (ab) replaced by a set of contacts. The contacts have started to open and there is an arc between them.

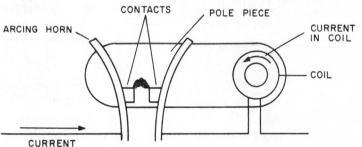

Fig. 5-6 Section of blowout magnet with straight conductor replaced
by a set of contacts. An arc is conducting between the contacts.

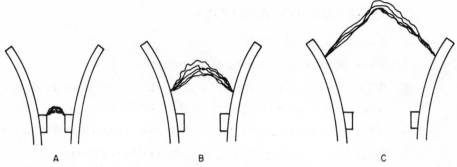

Fig. 5-7 Arc deflection between contacts.

Figure 5-7A shows the beginning deflection of the arc because of the effect of the motor action.

Figure 5-7B shows that the contacts are separated more than in figure 5-7A and the arc is beginning to climb up the horns because of the motor action and the effect of increased temperature.

Figure 5-7C shows the arc near the tips of the horns. At this point, the arc is so lengthened that it will be extinguished.

The function of the blowout magnet is to move the arc upward at the same time that the contacts are opening. As a result the arc is lengthened at a faster rate than will normally occur due to the opening of the contacts alone. It is evident that the shorter the time the arc is allowed to exist, the less damage it will do to the contacts.

AC MECHANICALLY HELD CONTACTORS AND RELAYS

Ac mechanically held contactors and relays are electro-mechanical devices that provide a safe and convenient means of switching circuits where quiet operation and continuity of circuit connection are requirements of the installation. For example, circuit continuity during power failures is often important in automatic processing equipment where a sequence of operation must continue from the point of interruption after power is resumed rather than return to the beginning of the sequence. Quiet operation of contactors and relays is required in many control systems used in hospitals, schools, and office buildings. Mechanically held contactors and relays generally are used in locations where the slight hum characteristic of alternating-current magnetic devices is objectionable.

In addition, mechanically held relays are often used in machine tool control circuits. These relays can be latched and unlatched through the operation of limit switches, timing relays, starter interlocks, other control relays, or pushbuttons. Generally, mechanically held relays are available in 10- and 15-ampere sizes; mechanically held contactors are available in sizes ranging from 30 amperes to 300 amperes.

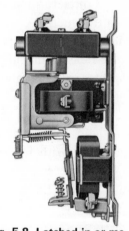

Fig. 5-8 Latched-in or mechanically held relay. The upper coil is energized momentarily to close contacts, and the lower coil is energized momentarily to open the contact circuit. The momentary energizing of the coil is an energy-saving feature. (Courtesy Square D Co.)

STUDY/DISCUSSION QUESTIONS

1. What are control relays?

2. What are typical uses for control relays?

3. Why are contactors described as both control pilot devices and large magnetic switches controlled by pilot devices?

4. What is the principal difference between a contactor and a motor starter?

5. What causes the arc to move upward in a blowout magnet?

6. Why is it desirable to extinguish the arc as quickly as possible?

7. What will happen if the terminals of the blowout coil are reversed?

8. Why will the blowout coil also operate on ac?

9. Why are mechanically held relays and contactors essential?

10. Why are mechanically held relays energy-saving devices?

Unit 6 Timing Relays

OBJECTIVES

After studying this unit, the student will be able to

- Identify the various types of timing relays.
- Explain the basic steps in the operation of timing relays.
- List the factors which will affect the selection of a timing relay for a particular use.
- List applications of several types of timing relays.
- Draw simple circuit diagrams using timing relays.

Many industrial control applications require timing relays that provide dependable service and are easily adjustable over the timing ranges. The proper selection and use of timing relays for a particular application can be made only after a study of the service requirements and with a knowledge of the operating characteristics inherent in each available device. A number of timing devices are available with features suitable for a wide variety of applications.

FLUID DASHPOT TIMING RELAYS

Magnetically operated, oil dashpot-type timing relays, figure 6-1, may be used on voltages up to 600 volts ac or dc. The contacts are operated by the movement of an iron core. The magnetic field of a solenoid coil lifts the iron core against the retarding force of a piston moving in an oil-filled dashpot. This type of relay is not very accurate and the piston must be allowed to settle back down to the bottom of the dashpot between successive timing periods. If the piston is not allowed to make a full return, the timing is very erratic. The dashpot timing relay provides time delay after the magnet is energized, and the contact may be either normally open or closed for different applications.

Unlike the magnetic overload relay, the dashpot timing relay operates with a potential coil connected across the line through contacts or switches. The overload relay operates with a current coil that is affected by the motor current load.

Fluid dashpot timing relays are used to control the accelerating contactors of motor starters, to time the closing or opening of valves on refrigeration equipment, or for any application where the operating sequence requires a delay. However, it is not necessary that the elapsed time of the delay be extremely accurate.

The use of silicone dashpot fluid (not an oil) in these relays helps to eliminate the effect of varying viscosity on the timing due to changes in ambient temperature. The silicone fluid operates successfully in an ambient temperature range

Fig. 6-1 Fluid dashpot timing relay. (Courtesy Square D Co.)

34

of +120° F. to –30° F. The timing range can be adjusted easily from two seconds to 30 seconds.

Multicontact dashpot timing relays are used for dc motor starting. When the coil is energized on this type of timing relay, the contacts close in succession with a time lag between each.

PNEUMATIC TIMERS

The construction and performance features of the pneumatic timer make it suitable for the majority of industrial applications. Pneumatic timers are unaffected by normal variations in ambient temperature or atmospheric pressure, are adjustable over a wide range of timing periods, have good repeat accuracy, and are available with a variety of contact and timing arrangements.

This type of relay has a pneumatic time-delay unit that is mechanically operated by a magnet structure. The time-delay function depends upon the transfer of air through a restricted orifice by the use of a reinforced synthetic rubber bellows or diaphragm. The timing range is adjusted by positioning a needle valve to vary the amount of orifice restriction.

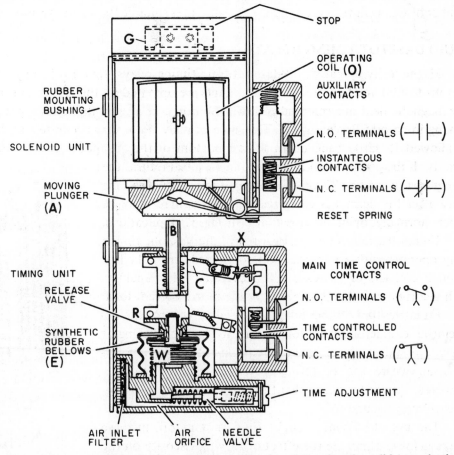

Fig. 6-2 Cross section of an ac "on-delay" timer, which provides time delay after the coil is energized. It is shown with the coil energized and the timer timed out. Schematic wiring symbols are shown (deenergized positions) for various portions of the timer. (Courtesy Allen-Bradley Co.)

The process of energizing or deenergizing pneumatic timing relays can be controlled by pilot devices such as pushbuttons, limit switches, or thermostatic relays. Since the power drawn by a timing relay coil is small, sensitive control devices may be used to control the operating sequence.

Pneumatic timing relays are used for motor acceleration and in automatic control circuits. Automatic control is necessary in applications where repetitive accuracy is required, such as controls for machine tools and control of sequence operations, industrial process operation, and conveyor lines.

Pneumatic timers provide time delay through two arrangements. The first, *on-delay*, means that the relay provides time delay when it is energized; the second arrangement, *off-delay*, means the relay is deenergized when it provides time delay. Figure 6-2 illustrates the operation of a pneumatic timer.

In the on-delay arrangement, when the operating coil (O) is energized, solenoid action raises the solenoid plunger (A). As a result, the pressure on rod (B) is released. Thus, spring (W), located inside the bellows (E), is allowed to push rod (B) upward. As (B) moves upward, it causes the off-center link mechanism (C) to move the end of the snap action mechanism (X) upward, which in turn raises (D) to operate the switch.

The speed with which the bellows rises is set by the position of the needle valve. The setting of the needle valve determines the time interval which must elapse between the solenoid closing and the rise of the bellows to operate the switch. If the needle valve is almost closed, an appreciable length of time is required for air to pass the valve and cause the bellows to rise.

When the solenoid is deenergized, the plunger (A) drops by force of gravity and by the action of the reset spring. The downward movement of the plunger forces (B) down, thus resetting the timer almost instantaneously.

In the off-delay arrangement, the solenoid in figure 6-2 is rotated 180°. Thus, when the operating coil (O) is energized, plunger (A) is held down, and (G) pushes against (B), holding the bellows (E) in a fully compressed position.

When the operating coil is deenergized, the reset spring forces (A) to rise. (G) also rises and releases the downward force on (B). The bellows (E) slowly expands, forcing (B) upward. As in the on-delay arrangement, as (B) moves up it trips the toggle, which in turn trips the switch.

MOTOR-DRIVEN TIMERS

When a process has a definite on and off operation, or a sequence of successive operations, a motor-driven timer is generally used. A typical application of a motor-driven timer is to control laundry washers where the loaded motor is run for a given period in one direction, reversed, and then run in the opposite direction. Motor-driven timers also are used where infrequent starting of large motors is required.

Generally, this type of timer consists of a small, synchronous motor driving a cam-dial assembly on a common shaft. A motor-driven timer successively closes and opens switch contacts which are wired in circuits to energize control relays or contactors to achieve desired operations.

Min. Time Delay: 1/20 second

Max. Time Delay: 3 minutes

Minimum Reset Time: .075 second

Accuracy: ±10 percent of setting

Contact Ratings:

Ac

6.0 Amp, 110 Volts
3.0 Amp, 220 Volts
1.5 Amp, 440 Volts
1.2 Amp, 550 Volts

Dc

1.0 Amp, 115 Volts
0.25 Amp, 230 Volts

Operating Coils: Coils can be supplied for voltages and frequencies up to 600 volts, 60 hertz ac and 250 volts dc.

Types of Contacts: One normally open and one normally closed. Cadmium silver alloy contacts.

Fig. 6-3 Typical specifications

Fig. 6-4 Motor-driven process timer in a general-purpose enclosure. (Courtesy Allen-Bradley Co.)

DC SERIES RELAY

Generally, the name of a relay is descriptive of its major purpose, construction, or principle of operation. For example, the coil of the dc series relay, figure 6-5, is connected in series with the starting resistance so that the starting current of the motor passes through it. The contacts of the series relay are connected to an auxiliary circuit.

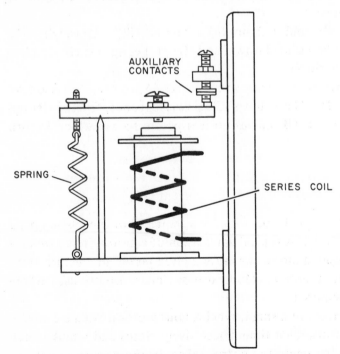

Fig. 6-5 Series type relay

**Fig. 6-6 Dc series relay
(Courtesy General Electric Co.)**

The relay contacts shown in figure 6-5 are connected to control the coils of a magnetic contactor. The armature is light and constructed so that it is very fast in operation. As the starting current passes through the coil, the armature is pulled down (overcoming the resistance of a spring), causing small contacts to open. When the current in the coil has decreased to a predetermined value, the spring pulls the armature back and the contacts close. The value of current at which the coil loses control of the armature is determined by the spring setting.

A common application of dc series relays is to time the acceleration of dc motors.

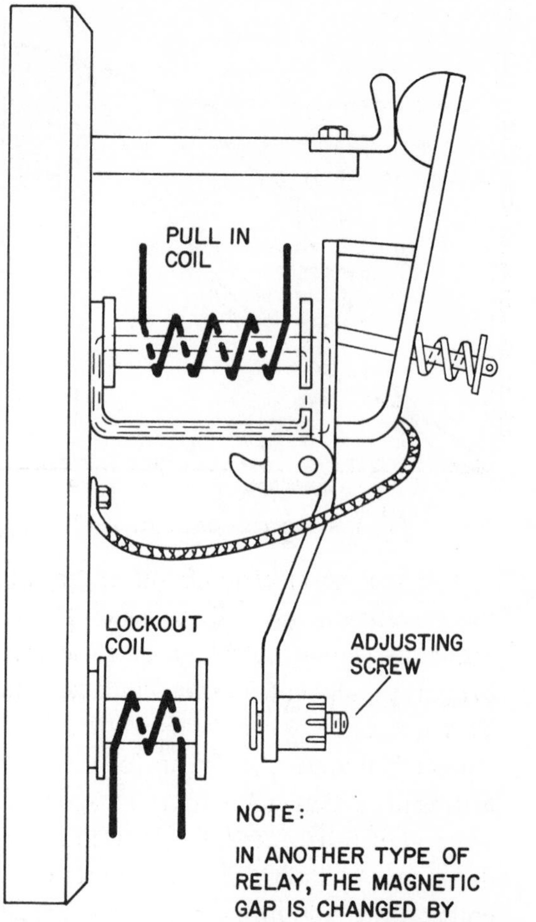

DC SERIES MAGNETIC LOCKOUT CONTACTOR

The dc series lockout relay consists of two coils connected in series. One coil, the lockout coil, acts to hold the contactor open while the other coil, the pull-in coil, acts to close the contactor. The pull-in coil is almost saturated at low-current values.

The current through the pull-in coil is the same as the current through the lockout coil. As a result, as the magnetic pull of one coil changes, the pull on the other coil will change

PULL IN COIL

LOCKOUT COIL

ADJUSTING SCREW

NOTE:

IN ANOTHER TYPE OF RELAY, THE MAGNETIC GAP IS CHANGED BY SHIMS LOCATED IN BACK OF THE COIL INSTEAD OF BY THE ADJUSTING SCREW

Fig. 6-7 Dc series magnetic lockout contactor.

also. However, since the closing (pull-in) coil has a relatively large amount of iron in its magnetic circuit, the current-torque curve of the closing coil is represented by curve C of figure 6-8. Due to the small amount of iron in the magnetic circuit of the lockout coil, the current-torque curve of the lockout magnet is represented by curve L of figure 6-8.

If the value of the motor-starting current is larger than ob, then the torque of the lockout coil predominates. When the motor accelerates and the current falls below the value ob, then the lockout coil loses control and the torque of the closing coil gains control.

MAGNETIC TIME LIMIT RELAY

If current is increasing in a coil, then the emf due to self-induction acts in a direction to oppose the increase of current in that coil. If a current is decreasing in a coil, the emf due to self-induction in the coil acts in a direction to oppose the decrease of current in the coil.

One type of magnetic time limit relay has a single coil wound on a hollow copper cylinder which contains an iron core. Other time limit relays have copper-jacketed coils. For either type of relay, time delay is provided when the relay drops out.

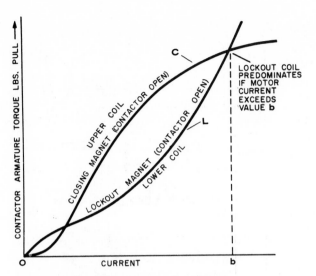

Fig. 6-8 Current-torque curves

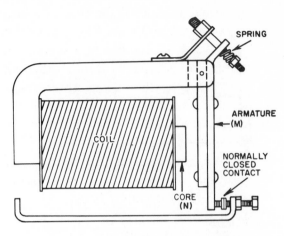

Fig. 6-9 Dc magnetic time limit relay.

When the electric circuit in the coil is broken, the current quickly falls to zero. Therefore, as the flux in the coil decreases, it cuts the short-circuited copper cylinder and induces a voltage in the cylinder. This induced voltage sends a current through the copper cylinder. As a result of this current, a flux is produced that holds up the magnet armature for a period of time after the coil circuit is broken. The delay time is limited by the number of turns on the coil and the amount of iron in the circuit.

Figure 6-9 is a sketch of a magnetic time limit relay. When the relay is energized, the armature (M) is drawn against the core (N). At the same time, the tension of the spring tends to draw the armature away from the core. As the flux maintained by the current in the copper sleeve dies away, the time at which the armature is released depends to some

Fig. 6-10 Time-delay contactor. (Courtesy General Electric Co.)

degree on the tension of the spring. The release time also can be varied by inserting a bronze shim in the gap between M and N; the thicker the shim, the lower the flux, and the sooner the armature is released.

The magnetic time limit relay is used to short out resistance steps in the startup of dc motors. With this type of relay armature pickup is instantaneous. The time delay at dropout is obtained by the use of a nonmagnetic armature shim and the adjustment spring.

CAPACITOR TIME LIMIT RELAY

If a capacitor is charged by momentarily connecting it across a dc line and then the capacitor dc is discharged through a relay coil, the current induced in the coil will decay slowly, depending on the relative values of capacitance, inductance, and resistance in the discharge circuit.

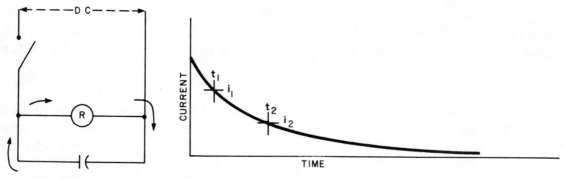

Fig. 6-11 Charged capacitor discharging through a relay coil. The graph at the right illustrates the current decrease in the coil.

If a relay coil and a capacitor are connected in parallel to a dc line, figure 6-11, the capacitor is charged to the value of the line voltage and a current appears in the coil. If the coil and capacitor combination is now removed from the line, the current in the coil will start to decrease along the curve shown in figure 6-11.

If the relay is adjusted so that the armature is released at current i_1, a time delay of t_1 is obtained. The time delay can be increased to a value of t_2 by adjusting the relay so that the armature will not be released until the current is reduced to a value of i_2.

The relay can be set by changing the spring tension and by using bronze shims as on the magnetic time limit relay. A potentiometer is sometimes used as an adjustable resistor to vary the time.

SELECTING A TIMING RELAY

In selecting a timing relay for a specific application, the Allen-Bradley Company recommends that the following factors be considered carefully.

- Length of time delay required.
- Timing range required.
- Allowable error.
- Cycle of operation and reset time.
- Cost.
- Additional requirements.

Length of Time Delay Required

The length of time delay required is determined by the type of machine or process that the timer will control. This time delay will range from a fraction of a second to as much as several minutes.

Timing Range Required

The phrase *timing range* means the various time intervals over which the timer can be adjusted. Timers are available which can be set for a time delay of 1 second, 100 seconds, or any value of delay between 1 and 100 seconds. When selecting a timer for use with a machine or process, the range of the timer should be wide enough to handle the various time-delay periods that may be required by the machine or process.

The exact timing value for any position within the timing range must be found by trial and error. A scale provided with a timer is intended primarily to permit the timer to be reset quickly to the timing position previously determined to be correct for a given operation.

Allowable Error

All timers are subject to some error; that is, there may be a time variation (plus or minus) between successive timing operations for the same setting. The amount of error varies with the type of timer and the operating conditions and is usually stated as some percentage of the time setting.

The percentage of error for any timer depends on the type of timer, the ambient temperature (especially at low temperatures), coil temperature, line voltage, and the length of time between operations.

Cycle of Operation Required and Reset Time

There are two basic types of timers. For one type of timer, the timer becomes operative when an electrical circuit opens or closes. A time delay then occurs before the application process begins. As soon as the particular process action is complete, the timer circuit resets itself. The circuit must be energized or deenergized each time the timing action is desired.

 The second type of timer is called a process timer. When connected into a circuit, this timer provides control for a sequence of events, one after another. The cycle is repeated continuously until the circuit is deenergized.

An important consideration in the selection of a timer is the speed at which the timer resets. *Reset time* is the time required for the relay mechanism to return to its original position. Some industrial processes require that the relay reset instantaneously, while other processes require a slow reset time. The reset time varies with the type of timing relay and the length of the time delay.

Cost

When there are several timers which all meet the requirements of a given application, then it is advisable to select the timer with the smallest number of operating parts. In other words, select the simplest timer. In most cases, this timer will probably be the least costly timer.

Additional Requirements

Several additional factors that must be considered in selecting a timer are:

- Type of power supply.
- Contact ratings.
- Timer contacts — a choice of normally open or normally closed contacts usually can be made.
- Temperature range — the accuracy of some timers varies with the temperature; however, it is temperatures below freezing that usually affect the timer accuracy.
- Dimensions — the amount of space available may have a bearing on the selection of a timer.

STUDY/DISCUSSION QUESTIONS

1. Which timer described in the unit is the most erratic? How may its timing period be affected by varying ambient temperatures?

2. List several applications for a motor-driven timer.

3. Which timer is the most commonly used in industrial applications?

4. How is the pneumatic timer adjusted?

5. List six factors to be considered when selecting a timing relay for a particular use.

6. List several additional factors which will affect the selection of a timing relay.

7. Draw an elementary diagram of a fractional horsepower, manual motor starter. When the starter contact is closed it will energize a pneumatic timing relay coil through the overload heater. After the timer coil is energized, the circuit will show a delayed closing contact that energizes a small motor on 120 volts.

Unit 7 Pressure Switches and Regulators

OBJECTIVES

After studying this unit, the student will be able to

- Describe how pressure switches, vacuum switches, and pressure regulators are used to control motors.
- List the adjustments which can be made to pressure switches.

Any industrial application which has a pressure sensing requirement can use a pressure switch. A large variety of pressure switches are available to cover the wide range of control requirements of pneumatic or hydraulic machines such as welding equipment, machine tools, high pressure lubricating systems, and motor-driven pumps and air compressors.

The pressure ranges over which pressure switches can maintain control also vary widely. For example, a diaphragm-actuated switch can be used where it is necessary to have a sensitive response to small pressure changes at low pressure ranges. A metal bellows-actuated control is used for pressures up to 2,000 pounds per square inch, while piston-operated hydraulic switches are suitable for pressures up to 15,000 psi.

The most commonly used pressure switches are single-pole switches, but two-pole switches are also used. Field adjustments of the range and the differential pressure (difference between the cut-in and cutout pressures) can be made for most pressure switches.

Pressure regulators provide accurate control of pressure or vacuum conditions for systems. Used as pilot control devices with magnetic starters, pressure regulators can control the operation of pump or compressor motors in a manner similar to that of pressure switches. Reverse action regulators can be used on pressure system interlocks to prevent the start of an operation until the pressure in the system has reached the desired level.

Pressure regulators consist of a Bourdon-type pressure gage and a control relay. Delicate contacts on the gage energize the relay and cause it to open or close. The relay contacts, then, are used to control a large motor starter to avoid damage to or burning of the gage contacts. Standard regulators will open a circuit at high pressure and close it at low pressure. Special reverse operation regulators will close the circuit at high pressure and open the circuit at low pressure.

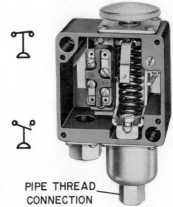

PIPE THREAD CONNECTION

Fig. 7-1 Industrial pressure switch with cover removed. Note operating knob. Also note wiring diagram symbols for normally closed and normally open contacts. (Courtesy Square D Co.)

STUDY/DISCUSSION QUESTION

1. Describe how pressure switches are connected to start and stop (a) small motors, and (b) large motors.

Unit 8 Float Switches

OBJECTIVES

After studying this unit, the student will be able to

- Describe the operation of float switches.

- List the sequence of operation for sump pumping or tank filling.

Float switches are designed to provide automatic control of ac and dc pump motor magnetic starters and automatic direct control of light motor loads.

The operation of a float switch is controlled by the upward or downward movement of a float placed in a water tank. The float movement causes a rod or chain and counterweight assembly to open or close electrical contacts. The float switch contacts may be either normally open or normally closed and may not be submerged. Float switches may be connected to a pump motor for tank or sump pumping operations or tank filling, depending on the contact arrangement.

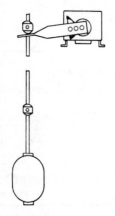

Fig. 8-1 Rod-operated float switch.

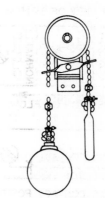

Fig. 8-2 Chain-operated float switch.

STUDY/DISCUSSION QUESTION

1. Describe the sequence of operations required to (a) pump sumps, and (b) fill tanks.

Unit 9 Flow Switches

OBJECTIVES

After studying this unit, the student will be able to

- Describe the purpose and functions of flow switches.
- Connect a flow switch to other electrical devices.
- Draw and read wiring diagrams of systems using flow switches.

A flow switch is a device which can be inserted in a pipe so that when liquid or air flows against a part of the device called a paddle, a switch is activated and either closes or opens a set of electrical contacts. These contacts may be connected to energize motor starter coils, relays, or indicating lights.

In general, a flow switch contains both normally open and normally closed electrical contacts, figure 9-1.

TOP VIEW OF SWITCH

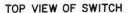

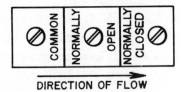

DIRECTION OF FLOW

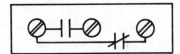

NO FLOW CONTACT POSITION

Fig. 9-1 Electrical terminals and contact arrangement of a flow switch.

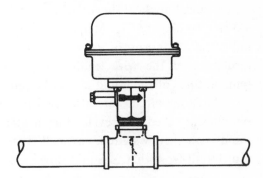

Fig. 9-2 Flow switch installed.

SIGNAL OR ALARM

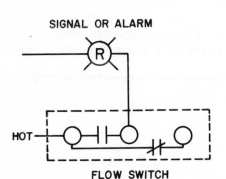

FLOW SWITCH

Fig. 9-3 Flow switch used to sound alarm or light signal when flow occurs.

SIGNAL OR ALARM

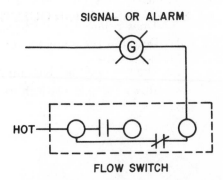

FLOW SWITCH

Fig. 9-4 Flow switch used to sound alarm or light signal when there is no flow.

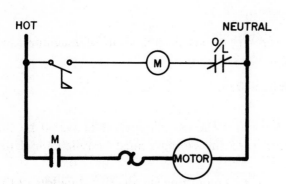

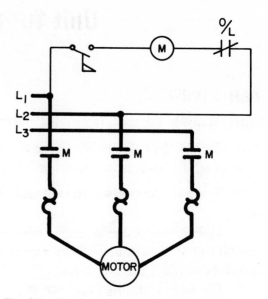

Fig. 9-5 Flow switch used with single-phase circuit; starts motor when flow occurs, stops motor when there is no flow.

Fig. 9-6 Flow switch used with three-phase circuit; starts motor when flow occurs, stops motor when there is no flow.

Figure 9-2 shows a flow switch installed in a pipe line tee. Half couplings are welded into larger pipes for flow switch installations.

Typical applications of flow switches are shown in figures 9-3 through 9-6 (see figures 9-3 and 9-4 on page 45). Many of these applications are found in the chemical and petroleum industries. Vaporproof electrical connections must be used with vaporproof switches. The insulation of the wire leading to the switches must be adequate to withstand the high temperatures of the liquid inside the pipe.

Airflow switches are also used in ducts in air conditioning systems. In addition, these switches are used to prevent duct heaters from energizing when there is no air movement in the duct. Airflow switches are called *sail* switches. While the construction of these switches is different from that of liquid flow switches, the electrical connections are similar.

STUDY/DISCUSSION QUESTIONS

1. What are typical uses of flow switches?

2. Draw a line diagram to show that when liquid flow occurs a green light will glow.

3. Draw a one-line diagram showing a bell that will ring in the absence of flow. Include a switch to turn off the bell manually.

Unit 10 Limit Switches

OBJECTIVES

After studying this unit, the student will be able to

- Explain how limit switches are used for the automatic operation of machines and machine tools.

- Wire a simple two-wire circuit using a limit switch.

The automatic operation of machinery requires the use of switches activated by the motion of the machinery. The repeat accuracy of the switches must be reliable and the response virtually instantaneous.

The size, operating force, stroke, and manner of mounting are all critical factors in the installation of limit switches due to mechanical limitations in the machinery. The electrical ratings of the switches must be carefully matched to the loads to be controlled.

In general, the operation of a limit switch begins when the moving machine or moving part of a machine strikes an operating lever which actuates the switch. The limit switch, in turn, affects the electrical circuit controlling the machine.

Limit switches are used as pilot devices in the control circuits of magnetic starters to start, stop, or reverse electric motors. Limit switches may be used either as control devices for regular operation or as emergency switches to prevent the improper functioning of machinery.

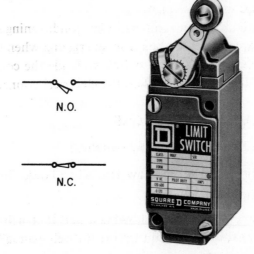

Fig. 10-1 Limit switch shown with electrical contact switches (Courtesy Square D Co.)

STUDY/DISCUSSION QUESTION

1. Draw a simple circuit showing a red pilot light energized when a limit switch is operated by a moving object.

Unit 11 Phase Failure Relays

OBJECTIVES

After studying this unit, the student will be able to

- Explain the purpose of phase failure relays for use with power supplies.
- List the hazards of phase failure and phase reversal.

If two phases of the supply to a three-phase induction motor are interchanged, the motor will reverse its direction of rotation. This action is called *phase reversal*. In the operation of elevators and in many industrial applications, phase reversal may result in serious damage to the equipment and injury to people using the equipment. In other situations, if a fuse blows or a wire to a motor breaks while the motor is running, the motor will continue to operate on single phase but will experience serious overheating. To protect motors against these conditions, phase failure and reversal relays are used.

One type of phase failure relay uses coils connected to two lines of the three-phase supply. The currents in these coils set up a rotating magnetic field that tends to turn a copper disc clockwise. This clockwise torque actually is the result of two torques, one polyphase torque which tends to turn the disc clockwise, and one single-phase torque which tends to turn the disc counterclockwise. The disc is kept from turning in the clockwise direction by a projection resting against a stop. However, if the disc begins to rotate in the counterclockwise direction, the projecting arm will move a toggle mechanism to open the line contactors and remove the motor from the line. In other words, if one line is opened, the polyphase torque disappears and the remaining single-phase torque rotates the disc counterclockwise and removes the motor from the line. In case of phase reversal, the poly-phase torque helps the single-phase torque turn the disc counterclockwise, and again, the motor is disconnected from the line.

Other designs of phase failure and phase reversal relays are available to protect motors, machines, and personnel from the hazards of open phase or reverse phase conditions. Some of these relays are more complex than the example given in the previous paragraph. For example, one type of relay consists of a static, current-sensitive network connected in series with the line and a switching relay connected in the coil circuit of the starter. The sensing network continuously monitors the motor line currents; if one phase opens, the sensing network immediately detects it and causes the relay to open the starter coil circuit to disconnect the motor from the line. A built-in delay of five cycles prevents nuisance dropouts caused by fluctuating line voltages.

STUDY/DISCUSSION QUESTIONS

1. What is the purpose of phase failure relays?
2. What are the hazards of phase failure and phase reversal?

Unit 12 Solenoid Valves

OBJECTIVES

After studying this unit, the student will be able to

- Describe the purpose and operation of a two-way and a four-way solenoid valve.
- Connect and troubleshoot solenoid valves.

Valves are mechanical devices designed to control the flow of fluids, such as oil, water, air, and other gases. While many valves are manually operated, electrically operated valves are most often used in industry because they can be placed close to the devices they operate, thus minimizing the amount of piping required. Remote control is possible by running a single pair of control wires between the valve and a control device such as a manually operated switch or an automatic device.

A solenoid valve is a combination of two basic units – a solenoid (electromagnet) and plunger (core) assembly, and a valve containing an orifice in which a disc or plug is positioned to regulate the flow. The valve is opened or closed by the movement of the magnetic plunger which is drawn into the solenoid when the coil is energized. The valve operates when current is applied to the solenoid. The valve returns automatically to its original position when the current ceases.

Most control pilot devices operate a single-pole switch, contact, or solenoid coil. The wiring diagrams of these devices are not difficult to understand and the actual devices can be connected easily into systems. It is recommended that the electrician know the purpose of and understand the action of the total industrial system for which various electrical control elements are to be used. In this way, the electrician will find it easier to design or assist in designing the electrical control system. In addition, he will find it easier to install and maintain the control system.

TWO-WAY SOLENOID VALVES

Two-way (in and out) solenoid valves, figure 12-1, page 50, are magnetically-operated valves which are used to control the flow of Freon, methyl chloride, sulphur dioxide, and other liquids in refrigeration and air conditioning systems. These valves can also be used to control water, oil, and air flow.

Standard applications of solenoid valves generally require that the valve be mounted directly in line in the piping with the inlet and outlet connections directly opposite each other. Simplified valve mounting is possible with the use of a bottom outlet which eliminates elbows and bends. In the bottom outlet arrangement, the normal side outlet is closed with a standard pipe plug.

The valve body is a special brass forging which is carefully checked and tested to insure that there will be no seepage due to porosities. The armature, or plunger, is made from a high grade stainless steel. The effects of residual magnetism are eliminated by the use of a kickoff pin and spring which prevent the armature from sticking. A shading coil insures that

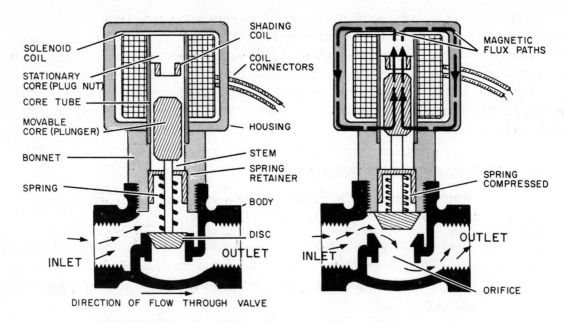

Fig. 12-1 Two-way solenoid valve. (Courtesy Automatic Switch Co.)

the armature will make a complete seal with the flat surface above it to eliminate noise and vibration.

It is possible to obtain dc coils with a special winding that will prevent the damage normally resulting from an instantaneous voltage surge when the circuit is broken. Capacitors are not required with this type of coil.

To insure that the valve will always seat properly it is recommended that strainers be used to prevent grit or dirt from lodging in the orifice or valve seat. Dirt in these locations will cause leakage. The inlet and outlet connections of the valve must not be reversed, because the tightness of the valve depends to a degree on pressure acting downward on the needle. This pressure is possible only with the inlet connected to the proper point as indicated on the valve.

FOUR-WAY SOLENOID VALVES

Electrically operated, four-port, four-way air valves are used to control a double-acting cylinder, figure 12-2.

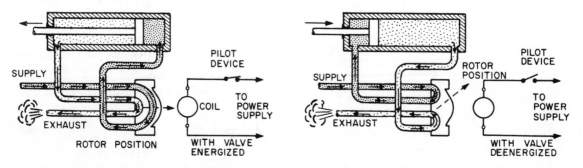

Fig. 12-2 Control of double-acting cylinder by a four-way, electrically operated valve.
(Courtesy Square D Co.)

When the coil is deenergized, one side of the piston is at atmospheric pressure, while the other side is acted upon by the line pressure. When the valve magnet coil is energized, the valve exhausts the high pressure side of the piston to atmospheric pressure. As a result, the piston and its associated load reciprocate in response to the valve movement.

Four-way valves are used extensively in industry to control the operation of the pneumatic cylinders used on spot welders, press clutches, machine and assembly jig clamps, tools, and lifts.

STUDY/DISCUSSION QUESTIONS

1. Why is it important to understand the purpose and action of the total operational system when working with controls?

2. If an electrically controlled, two-way solenoid valve is leaking, what is the probable cause?

3. What is the difference between a two-way solenoid valve and a flow switch?

Unit 13 Temperature Switches

OBJECTIVES

After studying this unit, the student will be able to

- Describe the operation of temperature switches.

- List several applications of temperature switches.

Temperature switches are designed to provide automatic control of temperature regulating equipment. Industrial temperature controllers are recommended for applications where the temperature to be controlled is higher than the normal or ambient temperature. In general, the applications of temperature control are more concerned with the temperature regulation of liquids than of gases. This is a result of the relatively greater conductivity between the bulb and a liquid as compared to the conductivity between the bulb and a gas (such as air). Thus, if air or gas temperatures are to be controlled, the sensitivity of the sensor decreases and the difference between the on and off points widens.

To operate a temperature switch, gas vapor or liquid pressure expands a metal bellows against the force of a spring. The expanding bellows moves an operating pin which then snaps a precision switch to its operating position when a preset point is reached (this action is similar to the operation of industrial pressure switches). The pressure which operates the switch is proportional to the temperature of a liquid in a closed bulb. The precision switch snaps back to its normal position when the pressure in the bellows drops enough to allow the main spring to compress the bellows.

There are many types of *thermostats* which can be used to provide automatic control of space heating and cooling equipment. A typical thermostat is a two-wire or greater control pilot device (switch) that is temperature actuated. Temperature-actuated switches are used to control circuits to operate heaters, blowers, fans, solenoid valves, pumps, and other devices.

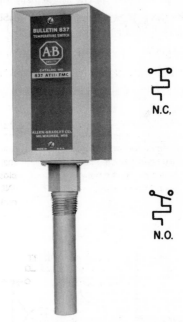

Fig. 13-1 Temperature control unit with vertical immersion bulb. The wiring symbols for normally open and normally closed contacts are shown to the right of the control unit. (Courtesy Allen-Bradley Co.)

STUDY/DISCUSSION QUESTION

1. What are some applications of temperature-actuated switches?

Section 3 Circuit Layout, Connections, and Symbols

Unit 14 Symbols

OBJECTIVES

After studying this unit, the student will be able to

- Identify common electrical symbols used in motor control diagrams.
- Use electrical symbols in drawing schematic and wiring diagrams.

If the directions for wiring electrical equipment are written without the use of diagrams, or if diagrams are used but must show each device as it actually appears, then the work and time involved both in preparing the directions and in installing the equipment will be very expensive. Therefore, it is common practice to use symbols to designate the various devices and components used in electrical systems. Because these symbols may not be similar to the physical appearance of the devices they represent, many symbols must be memorized.

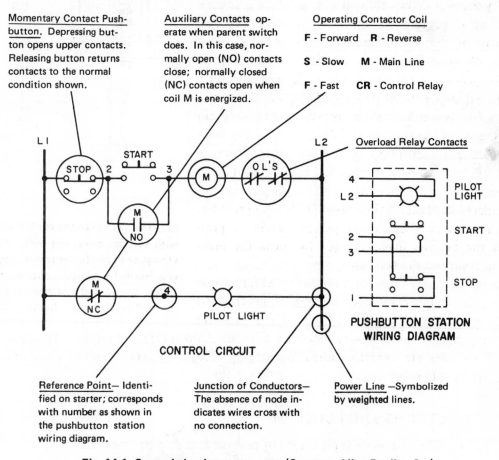

Momentary Contact Push-button. Depressing button opens upper contacts. Releasing button returns contacts to the normal condition shown.

Auxiliary Contacts operate when parent switch does. In this case, normally open (NO) contacts close; normally closed (NC) contacts open when coil M is energized.

Operating Contactor Coil

F - Forward   **R** - Reverse

S - Slow   **M** - Main Line

F - Fast   **CR** - Control Relay

Overload Relay Contacts

Reference Point— Identified on starter; corresponds with number as shown in the pushbutton station wiring diagram.

Junction of Conductors— The absence of node indicates wires cross with no connection.

Power Line —Symbolized by weighted lines.

Fig. 14-1 Control circuit components. (Courtesy Allen-Bradley Co.)

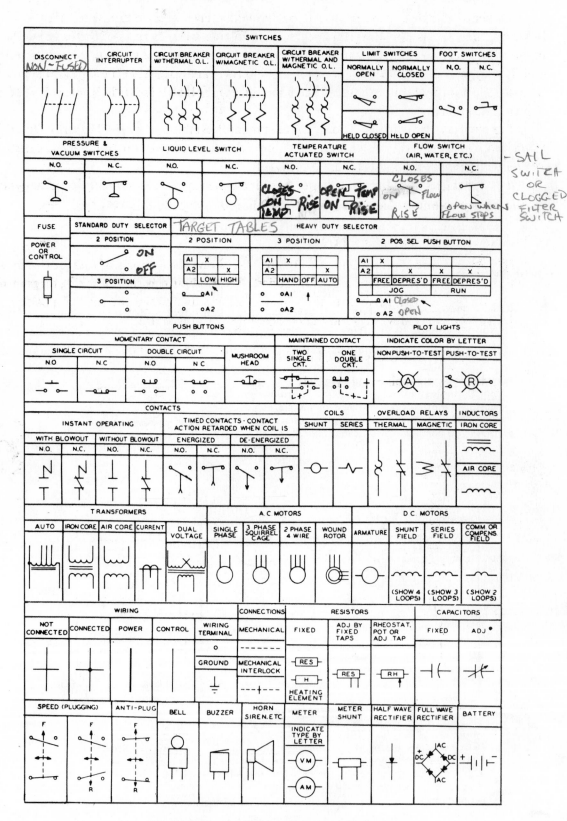

Fig. 14-2 Standard wiring diagram symbols.

LINE UP CLOSED
" DOWN OPEN

Standard symbols for electrical components have been established, figure 14-2. However, their use is not universal and many industries use variations of the basic symbols. In spite of the lack of standardization, the student should know the symbols presented in this unit as they will give him a firm basis for interpreting variations from the standard symbols. Of course, a thorough knowledge of the standard symbols will assist the student in reading electrical diagrams.

The symbols in figure 14-2 conform to standards established by the National Electrical Manufacturers Association (NEMA). Where NEMA standards do not exist, American Standards Association (ASA) standards are used.

It is recommended that the student refer frequently to the symbol list as he works through each until he is thoroughly familiar with all of the symbols given.

EXPLANATION OF COMMON SYMBOLS

The control circuit line diagram of figure 14-1 shows the symbols of each device in the circuit and indicates the function of each device. The pushbutton station wiring diagram on the right of figure 14-1 represents the physical station and shows the relative position of each device, the internal wiring, and the connections with the starter.

STUDY/DISCUSSION QUESTIONS

Identify the following symbols

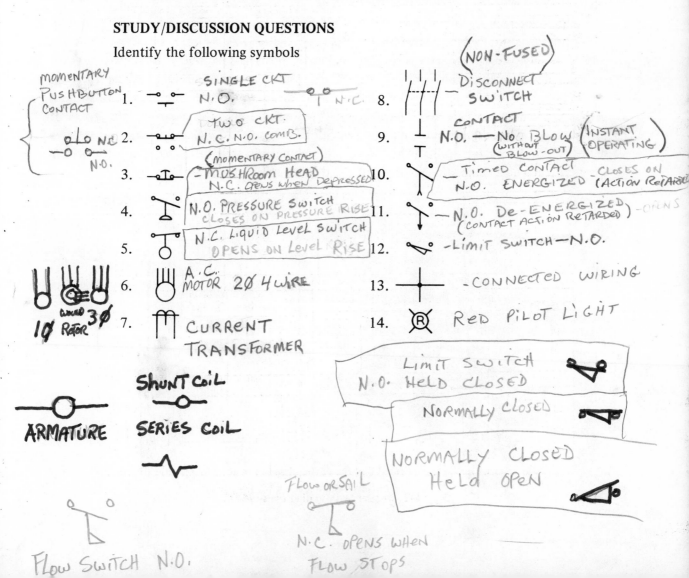

1. SINGLE CKT N.O.   N.C.
   MOMENTARY PUSHBUTTON CONTACT
2. TWO CKT. N.C. N.O. COMB.
   NC NO
3. MUSHROOM HEAD N.C. OPENS WHEN DEPRESSED
   (MOMENTARY CONTACT)
4. N.O. PRESSURE SWITCH CLOSES ON PRESSURE RISE
5. N.C. LIQUID LEVEL SWITCH OPENS ON LEVEL RISE
6. A.C. MOTOR 2∅ 4 WIRE
7. CURRENT TRANSFORMER
   1∅   WOUND 3∅ ROTOR

8. DISCONNECT SWITCH (NON-FUSED)
9. CONTACT N.O. — No. BLOW (WITHOUT BLOW-OUT) (INSTANT OPERATING)
10. Timed CONTACT N.O. ENERGIZED (ACTION RETARDED) CLOSES ON
11. N.O. DE-ENERGIZED (CONTACT ACTION RETARDED) OPENS
12. Limit SWITCH — N.O.
13. CONNECTED WIRING
14. RED PILOT LIGHT

ARMATURE

ShuNT Coil
SERiES Coil

Limit Switch N.O. HELD CLOSED
NORMALLY CLOSED
NORMALLY CLOSED Held OPEN

FLOW or SAIL
N.C. OPENS WHEN FLOW STOPS

FLOW SWITCH N.O.

Unit 15 The Interpretation and Application of Simple Wiring and Elementary Diagrams

OBJECTIVES

After studying this unit, the student will be able to

- Draw an elementary line diagram using a wiring diagram as a reference.
- Describe the differences between the advantages of two- and three-wire control.
- Connect motor starters with two- and three-wire control.
- Identify common terminal markings.
- Read and use target tables.
- Connect drum reversing controllers.
- Interpret motor nameplate information.
- Connect dual-voltage motors.

LINE DIAGRAMS AND WIRING DIAGRAMS

Most electrical circuits can be represented by two types of diagrams: a wiring diagram and a line diagram. Wiring diagrams include all of the devices in the system and show the physical relationships between the devices. All poles, terminals, contacts and coils are shown on each device. These diagrams are particularly useful in wiring circuits because the connections can be made exactly as they appear on the diagram, wire for line. A wiring diagram gives the necessary information for the actual wiring of a circuit and provides a means of physically tracing the wires for troubleshooting purposes or when normal preventative maintenance is necessary. In other words, the actual physical installation and the wiring diagram coincide as far as the locations of the devices and wiring are concerned.

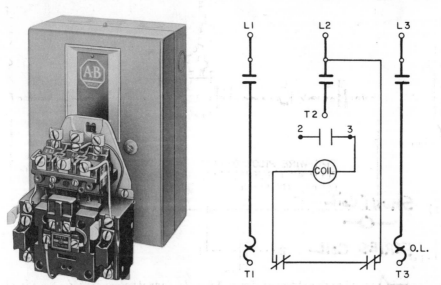

Fig. 15-1 A motor starter (left) and the wiring diagram of the same starter (right)
(Courtesy Allen-Bradley Co.)

56

In trying to determine the electrical sequence of the circuit, however, it is not easy to use the wiring diagram. For this reason, a rearrangement of the circuit elements is made to form what is called a *line diagram.* Line diagrams, also called *elementary diagrams* and *schematic diagrams,* are widely used in industry, and it is a decided advantage to the electrician to learn to use them.

The line diagram also represents the electrical circuit, but does so in the simplest manner possible. No attempt is made to show the various devices in their actual physical positions. All control devices are shown located between vertical lines that represent the source of control power. Circuits are shown connected as directly as possible from one of the source lines horizontally through contacts and current-consuming devices to the other source line. All connections are made so that the functions and sequence of operations of the various devices and circuits can be traced easily. The schematic diagram is invaluable in troubleshooting because it shows the effect that opening or closing various contacts has on other devices in the circuit. The schematic diagram attempts to convey as much information as possible with the least amount of confusion.

The student should study figure 15-1 to become familiar with a motor starter as it is represented by a wiring diagram. The principal parts of the starter are labeled on the diagram so that a comparison can be made with the actual starter. This study should help the student to visualize the starter when viewing a wiring diagram, and will help in making the correct connections when the starter is actually wired. Note that the wiring diagram shows as many parts as possible in their proper relative positions. It is not necessary to show the armature and crossbar or the overload reset mechanism of the starter in the wiring diagram since these parts are not involved in the actual wiring.

The wiring diagram in figure 15-1 represents one size of starter. Other starter sizes will have similar diagrams and will be wired in a similar manner although some of the connections may not have the same physical locations on the starters.

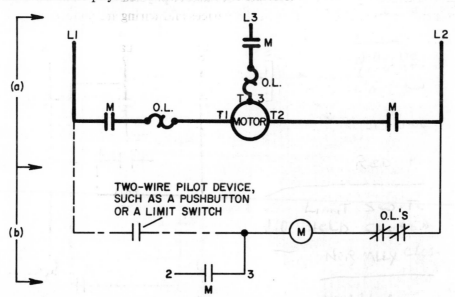

Fig. 15-2 Line diagram of starter shown in figure 15-1: (a) Line diagram showing the simplicity of the power circuit for the line voltage starter; (b) Completed control wiring elementary style. Broken lines and motor are external field wiring that completes the circuit.

Figure 15-2, page 57, is an elementary or line diagram of the starter shown in figure 15-1. In many cases, the line diagram does not show the power circuit to the motor because of its simplicity. Note that the line 1 (L1) circuit is completed to the motor (when the coil is energized) through main contact (M) and the overload heater to terminal 1 (T1) of the motor. All of the contacts close simultaneously when the coil is energized and the motor starts with the applied line voltage.

The control wiring of the starter is shown in solid lines, figure 15-2B. This completes the wiring of the actual starter. The external wiring completed by the electrician in the field is shown with broken lines. Note how the control wiring is rearranged to form one horizontal path from line 1 (L1) to line 2 (L2). This figure shows how the control wiring diagram is converted in appearance to a line diagram. Although the drawings differ, they are the same electrically. In general, it is easier to trace the operation of the starter by viewing the line drawing than it is to use the wiring diagram of the same starter.

TWO-WIRE CONTROL – Low Voltage Release

Two-wire control is also known as *no-voltage release* or *low-voltage release*. Two-wire control of a starter means that the starter drops out when there is a voltage failure and picks up as soon as the voltage returns. In figure 15-3, the pilot device is unaffected by the loss of voltage and its contact remains closed, ready to carry current as soon as line voltage returns to normal. This form of control is used frequently to operate supply and exhaust blowers and fans.

Two-wire control is also illustrated in schematic form in figure 15-2B. While two-wire control means that an operator does not have to be present to restart a machine following a voltage failure, this type of restart can be a safety hazard to personnel and machinery. For example, materials in process may be damaged due to the sudden restart of machines upon restoration of power. (Remember, power may return without warning.)

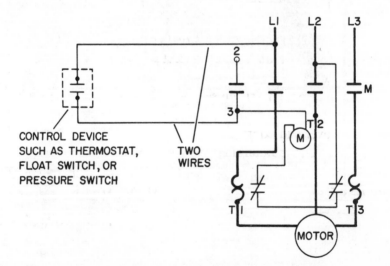

Fig. 15-3 Two wires lead from the pilot device to the starter. The phrases "no-voltage release" and "two-wire control" should indicate to the student that an automatic pilot device, such as a limit switch or float switch, opens and closes the control circuit through a single contact.

THREE-WIRE CONTROL — L.V. PROTECTION

Three-wire control, known as *no-voltage protection* or *low-voltage protection,* is also a basic and commonly used control circuit.

Three-wire control of a starter means that the starter will drop out when there is a voltage failure, but will not pick up automatically when voltage returns. The control circuit in figure 15-4 is completed through the "stop" button and also through a "holding" contact (2-3) on the starter. When the starter drops out, the holding contact opens and breaks the control circuit until the "start" button is pressed to restart the motor.

The main distinction between the two types of control is that in no-

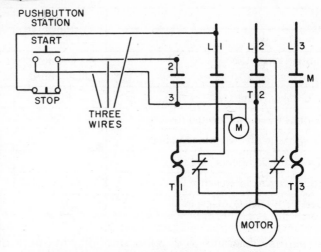

Fig. 15-4 Three wires lead from the pilot device to the starter. The phrases "no-voltage protection" and "three-wire control" should indicate to the student that the most common means of providing this type of control is a start-stop pushbutton station.

voltage release (two-wire control), the coil circuit is maintained through the pilot switch contacts; in no-voltage protection (three-wire control), the circuit is maintained through a "stop" contact on the pushbutton station and an auxiliary contact on the starter. The designations "two wire" and "three wire" are used only because they describe the simplest applications of the two types of control. Actually, for more complicated circuits, there may be more wires leading from the pilot device to the starter. However, the principle of two-wire or three-wire control will still apply in these situations.

Three-wire control prevents the restarting of machinery when power is restored. It will be necessary for an operator to press the "start" buttons to resume production when it is safe to do so.

It may have been noted in figure 15-2B that contact 2-3 is not included in the two-wire circuit. This is the *holding* or *maintaining* contact which is connected in a three-wire control circuit as shown in figure 15-4.

Referring to figure 15-4, it can be seen that when the start button is pressed, coil M is energized from line 1 (L1) through the normally closed stop pushbutton, the depressed start button, the coil, and overload control contacts to line 2 (L2). When coil M is energized, it closes all M contacts. The maintaining contact (2-3) holds the coil circuit when the start button is released. When the stop pushbutton is pressed, the stop contact opens and the coil is deenergized. As a result, all M contacts are opened and the motor stops. In the event of a motor overload, excessive current is drawn from the line, causing the thermal heaters to overheat. Thus, the normally closed overload relay control contacts are opened and the motor stops.

CONTROL AND POWER CONNECTIONS

The correct connections and component locations for line and wiring diagrams are indicated in Table 15-1. The student should compare the information given in the table with

Table 15-1 Control and Power Connections — 600 Volts or Less — Across-the-Line Starters

	1 Phase	3 Phase	Direct Current	
Line Markings	L1, L2	L1, L2, L3	L1, L2	
Ground When Used	L1 Always Ungrounded	L2	L1 Always Ungrounded	
O/L Relay Heaters in	L1	T1, T2, T3	L1	
Control Circuit Connected to	L1, L2	L1, L2	L1, L2	
Control Circuit Switching Connected to	L1	L1	L1	
Contactor Coil Connected to	L2	L2	L2	
For Reversing Interchange Lines	—	L1, L3	—	
Overload Relay Contacts in	L2	L2	L2	

actual line diagrams to develop an ability to interpret the table quickly and use it correctly. For example, refer to figure 15-4 and note that as indicated in the three-phase column of the table, the control circuit switching is connected to line 1 (L1) and the contactor coil is connected to line 2 (L2).

TARGET TABLES

The phrase, *target table,* or *sequence chart,* refers to a chart which lists the sequence of operation of identified contacts in a circuit. A target table is useful in interpreting circuit drawings and for simplifying troubleshooting.

One particular advantage of using a target table is that contacts on selector switches, figure 15-5, need not be drawn close together on the elementary diagram, but can be spaced over the diagram, if necessary. Dotted lines should be drawn on the diagram to connect the various contacts of one switch.

Selector Switch Diagramming

Figure 15-5 shows a typical, three-position selector switch (left), its line diagram symbol (center), and a target table (right) indicating the condition of each contact for each position of the selector switch. The "X" represents closed contacts. A small target table, such as that for a single two-circuit, double-contact switch, can be shown above the switch contact on the elementary diagram. Larger target tables should be shown to one side or below the line diagram.

The target table for the three-position selector switch, figure 15-5, shows that in the OFF position (arrow) no contacts are closed. In the HAND position, contact A1 is closed and A2 is open. In the AUTO (automatic) position, A2 is closed and A1 is open.

Drum Switch Diagramming

A drum switch consists of a set of moving contacts mounted on and insulated from a rotating shaft. The shaft also

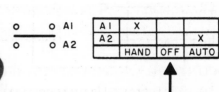

A1	X		
A2			X
	HAND	OFF	AUTO

Fig. 15-5 Heavy duty, three-position selector switch.

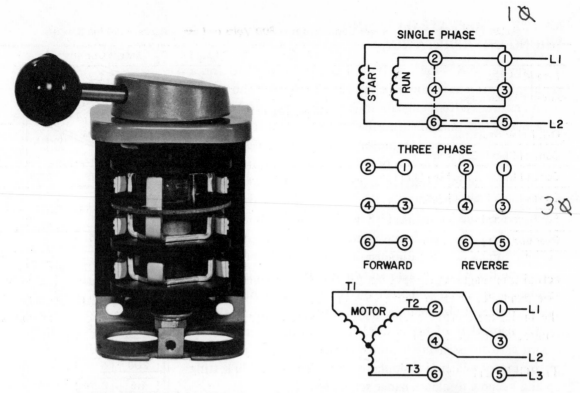

Fig. 15-6 Drum switch with cover removed; diagrams show switch connections to single-phase and three-phase motors.

has a set of stationary contacts that make and break contact with the moving contacts as the shaft is rotated, figure 15-6.

The term drum switch is applied to cam switches as well. Drum switches may be used to carry the starting current of a motor directly or may be used to handle only the current for the pilot devices that control the main motor current.

In figure 15-7A, the terminal markings (enclosed in dashed-line box) indicate how the line and load are connected to the drum switch. The target table, figure 15-7B, shows which contacts are closed to reverse the three-phase motor and which contacts are closed to cause the motor to run forward.

Target tables are often used in complicated circuits to simplify the interpretation of the operation of large multicircuit drum switches.

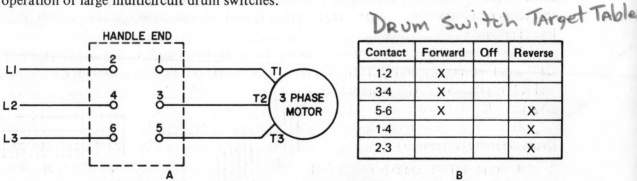

Contact	Forward	Off	Reverse
1-2	X		
3-4	X		
5-6	X		X
1-4			X
2-3			X

Fig. 15-7 A target table is used to assist in the interpretation of drum switch operation where a three-phase or a single-phase motor is reversed by interchanging two motor leads.

MOTOR NAMEPLATE DATA AND WIRING INTERPRETATION

Figure 15-8 shows a typical motor nameplate for a standard three-phase, nine-lead motor. Table 15-2 shows one manufacturer's letter designations for various types of motors. In general, each manufacturer has an individual set of letter designations. A type ML motor is indicated in figure 15-8. According to Table 15-2, a type **ML** motor has normal starting torque, normal starting current, and normal slip. A mechanical characteristic is that the motor is dripproof.

Also shown in figure 15-8 is the motor frame number which refers to the frame dimensions indicated on the manufacturer's detailed dimension sheets. Each motor has a serial number which is also shown on the nameplate. This number is recorded and kept on file by the manufacturer. The record card also includes most of the information shown on the nameplate of the motor. The manufacturer uses the record card to assist in customer service, while the purchaser of the motor will find a similar record card helpful in setting up and keeping to maintenance schedules.

The horsepower (h.p.) rating on the nameplate refers to the output of the motor, as measured by mechanical means.

The code letter on the nameplate indicates the locked rotor kva. input per horsepower as shown in **NEMA** standards. Duty indicates the duty cycle for which the motor is designed, such as 15 minutes, 30 minutes, or continuous operation. The degrees celsius rise above ambient temperature for a motor operating under full load is shown in the °C rise box. Rpm refers to the synchronous speed of the motor in revolutions per minute. The current in amperes drawn by the motor is listed under AMPS and varies as

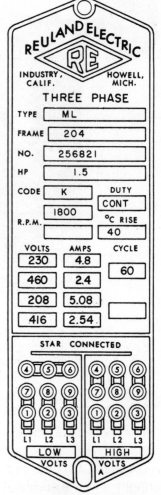

Fig. 15-8 Standard three-phase motor nameplate, nine leads.

indicated according to the voltage system to which the motor is connected. Motors draw half as much current on high voltages as they do on low voltages. However, the same amount of power is used for either high- or low-voltage connections. Low-current connections mean that smaller wire sizes, conduits, feeders, and switch gear can be used. If available, connection to high voltages will mean a financial savings in wiring materials and less wasted electrical energy due to decreased voltage drop in the circuits. The motor starter coil voltage must be equal to the supply voltage and the overload heaters must be rated for the proper current value.

At the bottom of the nameplate, figure 15-8, the LOW volts indication refers to motor terminal connections to 230 and 208 volts; the HIGH volts indication refers to 460- and 416-volt connections. For a high-voltage connection, the nameplate shows that motor lead terminals 4 and 7 are connected together, as are 5 and 8, and 6 and 9. The connections are insulated from each other in the motor terminal connection box. The three remaining leads

Table 15-2 Sample Nameplates and Connection Diagrams

Nameplate	Description	Connection
A	Dual Voltage	Three-Phase — Star or Wye
N	Dual Voltage	Three-phase — Delta
F	Dual Voltage	Single-phase

Type Designations

Type	Electrical Characteristics
ML	Normal starting torque and normal starting current. Normal slip.
HML	High starting torque — low starting current. Normal slip.
WML	Wound rotor — slip ring.
MLS	Capacitor start — induction run — single-phase.
RLS	Repulsion-induction start and run — single-phase.

	Mechanical Characteristics
ML	Dripproof
MLU	Splashproof
MLE	Totally enclosed nonventilated
MLF	Totally enclosed fan cooled
MLV	Vertical solid shaft
MLVH	Vertical hollow shaft
GML	Helical Parallel Shaft Motor Reducer
GMLW	Right Angle Worm Gear Motor Reducer
Fluid	Fluid Shaft Motors

(1, 2, and 3) are connected to three-phase supply lines L1, L2, and L3. Actually, this connection is to the load side of the starter terminals (T1, T2, and T3).

Figure 15-9 shows a recommended method of determining the three-phase motor terminal connections in the event that the motor nameplate is inaccessible, destroyed, or lost when the motor is installed.

Star or delta test equipment is used to determine the unknown internal wiring connections of the motor. A continuity test will find one group of three common leads. For example, such a test will establish the center of the star high-voltage connection (at T7, T8, and T9) in figure 15-9, page 64. There should be only one group of three motor leads common to each other for a star-connected motor, and three groups of three leads common to each other on a delta, internally-connected motor, (T1-T4-T9, T2-T5-T7, T3-T6-T8).

The designation "A" at the bottom of the nameplate in figure 15-8 is part of the manufacturer's information as shown in the column headed NAMEPLATE in Table 15-2.

The nameplates for some motors include a service factor reference. The service factor of a general-purpose motor is an allowable overload. The amount of allowable overload is indicated by a multiplier, which, when applied to the normal horsepower rating, indicates the permissible loading.

SINGLE-PHASE, DUAL-VOLTAGE MOTOR CONNECTIONS

The connections for dual-voltage, single-phase ac motors are shown in figure 15-10. For low-voltage operation, the stator coils are connected in parallel; for high-voltage operation, they are connected in series. The instantaneous current direction is indicated by

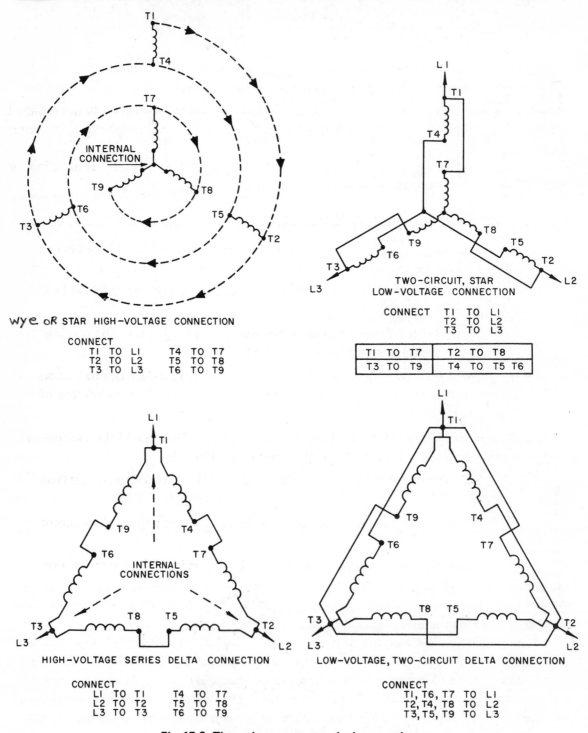

WYE OR STAR HIGH-VOLTAGE CONNECTION

```
CONNECT
  TI  TO  LI      T4  TO  T7
  T2  TO  L2      T5  TO  T8
  T3  TO  L3      T6  TO  T9
```

TWO-CIRCUIT, STAR
LOW-VOLTAGE CONNECTION

```
CONNECT   TI  TO  LI
          T2  TO  L2
          T3  TO  L3
```

TI TO T7	T2 TO T8
T3 TO T9	T4 TO T5 T6

HIGH-VOLTAGE SERIES DELTA CONNECTION

```
CONNECT
  LI  TO  TI      T4  TO  T7
  L2  TO  T2      T5  TO  T8
  L3  TO  T3      T6  TO  T9
```

LOW-VOLTAGE, TWO-CIRCUIT DELTA CONNECTION

```
CONNECT
  TI, T6, T7  TO  LI
  T2, T4, T8  TO  L2
  T3, T5, T9  TO  L3
```

Fig. 15-9 Three-phase motor terminal connections.

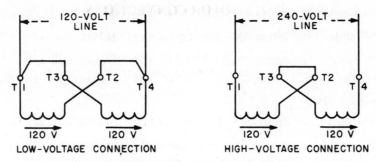

Fig. 15-10 Single-phase, dual-voltage motors.

arrows at the motor coils. In general, starting windings are factory connected to the terminal screws for 120 volts. This means that for the higher voltage value, the connections are center tapped.

All external wiring motor leads are located in the motor terminal connection or conduit box.

STUDY/DISCUSSION QUESTIONS

1. In which diagram do the physical locations of the wiring and the devices match the drawing?

2. Which diagram conveys as much information as possible with the least amount of confusion?

3. Why is it not necessary to show the armature and crossbar or the overload reset mechanism in a wiring diagram?

4. Which control wiring scheme can cause damage to industrial production or processing with the sudden restart of machines upon restoration of the power?

5. In developing line or wiring diagrams, in which lines should the overload relays be placed for three-phase motor starters?

6. In which line of a three-phase starter are the overload relay contacts placed?

7. The control circuit is connected to which lines for a three-phase motor starter?

8. Identify the terminal markings for the following delta-wound motor.

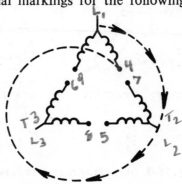

9. The contactor coil is connected to which line of a three-phase contactor?

10. What does the letter "X" mean on a target table?

11. A motor nameplate reads 220-440 volts, 30-15 amperes. When connected to a 440-volt supply, how much current will the motor draw fully loaded?

12. To what does the code letter on a motor nameplate refer?

13. Draw an elementary line diagram of the control circuit from the wiring diagram below. Exclude power or motor circuit wiring.

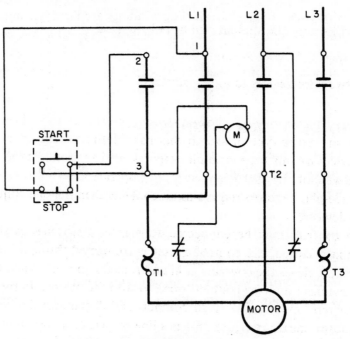

14. Draw a line diagram of the control circuit below.

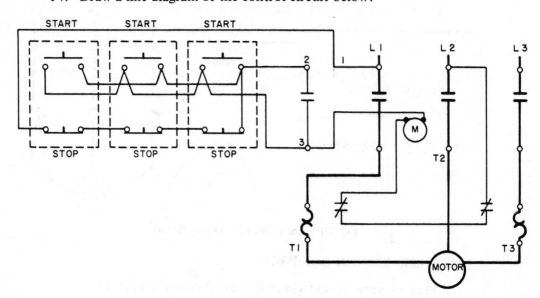

Section 4 Basic Control Circuits

Unit 16 Two-Wire Controls

OBJECTIVES

After studying this unit, the student will be able to

- List advantages of two-wire control.

- Connect two-wire devices to motor starters.

A two-wire control may be a toggle switch, pressure switch, float switch, limit switch, thermostat, or any other type of switch that has a definite on or off position. As indicated in unit 15, devices of this type generally are designed to handle small currents. Two-wire control devices will not carry sufficient current to operate large motors. In addition, 240-volt motors and three-phase motors require more contacts than the one contact usually provided with two-wire devices.

Two-wire controls may be connected to magnetic switches as shown in figure 16-1. When the switch is closed, the control circuit is completed through the coil (M). When the coil is energized, it closes the contacts at M and runs the motor. When the switch is opened, the coil is deenergized and the contacts open to stop the motor. In the case of an overload, the thermal heaters open the overload contacts and deenergize the coil, thus stopping the motor. The motor starter in figure 16-1 is a line voltage or across-the-line starter (described in unit 3).

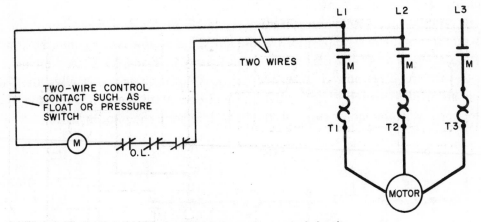

Fig. 16-1 Basic two-wire control circuit.

STUDY/DISCUSSION QUESTION

1. What are some advantages to the use of two-wire control?

Unit 17 Three-Wire and Separate Controls

OBJECTIVES

After studying this unit, the student will be able to

- Describe the basic sequence of operations for a three-wire control circuit.
- Describe the basic sequence of operations for a separate control circuit.
- List the advantages of each type of circuit.
- Connect three-wire and separate control circuits.

THREE-WIRE CONTROLS

Three-wire controls are devices such as momentary contact pushbuttons (stop-start stations) and double-acting thermostats.

In general, three-wire devices are connected as shown in figure 17-1. Although the arrangement of the various parts may vary from one manufacturer's switch to another, the basic circuit remains the same.

The sequence of operation for this circuit is as follows: when the start button is pushed, the circuit is completed through the coil (M), and the contacts at M close. When the start button is released, the holding contact at M keeps the coil energized. With the holding contact closed, the circuit is still complete through the coil. If the stop button is pushed, the circuit is broken, the coil loses its energy, and the contacts at M open. When the stop button is released, the circuit remains open because both the holding contact and the start button are open. The start button must be pushed again to complete the circuit. The operation of the overload protection will have the same effect; that is, if the supply voltage fails, the circuit is deenergized. When the supply voltage returns, the circuit remains open until the start button again is pushed. This arrangement is called *no-voltage protection* and protects both the operator and the equipment.

The pushbutton station wiring diagram, figure 17-1B, is a representation of the physical station and shows the relative positions of the units, the internal wiring, and the connections with the starter. For similar diagrams, review figures 14-1 and 15-1. A pilot light can be added to this circuit to indicate when the motor is not running. Normally closed auxiliary contacts are used to switch the pilot light on and off. With the motor running, these contacts are open; with the motor stopped, they are closed and the pilot light is lit. A pilot light

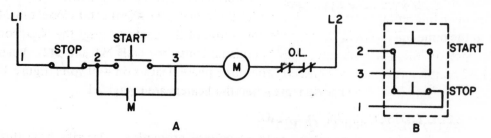

Fig. 17-1 Basic three-wire control circuit.

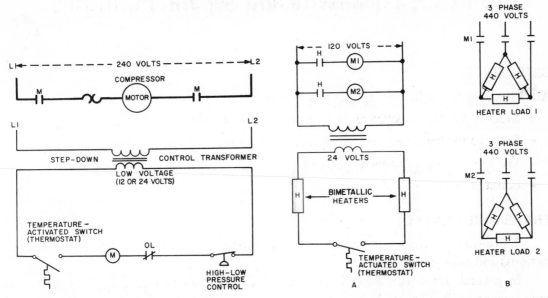

Fig. 17-2 Separate control used in air-conditioning cooling circuit.

Fig. 17-3 Separate control used in air-conditioning heating circuit.

may be installed to indicate when the motor is running. This is accomplished by connecting it between control Terminal 3 and Line 2. Except for this modification, the circuit is a basic three-wire, pushbutton control circuit.

SEPARATE CONTROL

If the control circuit is separated from the power circuit, the control connections are not made to L1 and L2. A separate source, such as an isolating transformer or an independent voltage supplies the power to the control circuit. This independent voltage is separate from the main power used for the motor.

One form of separate control is shown in figure 17-2. This is an elementary diagram of a cooling circuit for a commercial air-conditioning installation. When the thermostat calls for cooling, the compressor motor starter coil (M) is energized through the step-down transformer. When coil (M) is energized, power contacts (M) in the 240-volt circuit close to start the refrigeration compressor motor. Since the control circuit is separated from the power circuit by the isolating control transformer, there is no electrical connection between the two circuits.

Another example of separate control is the thermal relay shown in figure 17-3. This control consists of a line to a low-voltage transformer, two low-voltage heaters (H), and two normally open line voltage contacts. When heat is required, the thermostat closes the 24-volt circuit to the bimetallic heaters. In seconds, the action of the heaters causes the N.O. contacts (H) to close on the 120-volt side, thus energizing contactor coils M1 and M2. When the M1 and M2 contactors close, they energize the three-phase heaters on 440 volts, figure 17-3B. Circulating blowers (not shown) will start when the heaters are energized.

STUDY/DISCUSSION QUESTION

1. Draw a control circuit with no-voltage protection. Describe how this method of wiring protects the machine operator.

Unit 18 Hand-Off Automatic Controls

OBJECTIVES

After studying this unit, the student will be able to

- State the purpose of hand-off automatic controls. *(MANUAL)*
- Connect hand-off automatic controls. *(MANUAL)*
- Draw diagrams using hand-off automatic control. *(MANUAL)*

(MANUAL) Hand-off automatic switches are a form of selector switch used to select the function of a motor controller either manually or automatically. The switch may be a separate unit or built into the starter enclosure cover. A typical control circuit using a selector switch is shown in figure 18-1.

In the HAND (manual) position, coil (M) is energized all the time and the motor runs continuously. In the OFF position, the motor does not run at all. In the AUTO position, the motor runs whenever the two-wire control device is closed. The control device may be a pressure switch, limit switch, thermostat, or other two-wire control.

The heavy-duty, three-position selector switch shown in figure 18-2 is used for manual and automatic control.

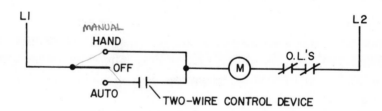

Fig. 18-1 Standard duty, three-position selector switch in control circuit.

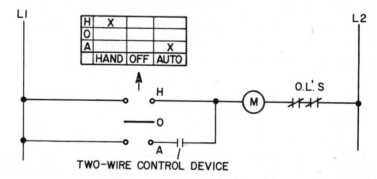

Fig. 18-2 Diagram of a heavy duty, three-position selector switch in control circuit.

STUDY/DISCUSSION QUESTIONS

1. A selector switch and two-wire pilot device can not be used to control a large motor directly, but rather must be connected to a magnetic starter. Explain why this is true.

70

2. Determine the minimum number of wires in each conduit shown in the following diagram.

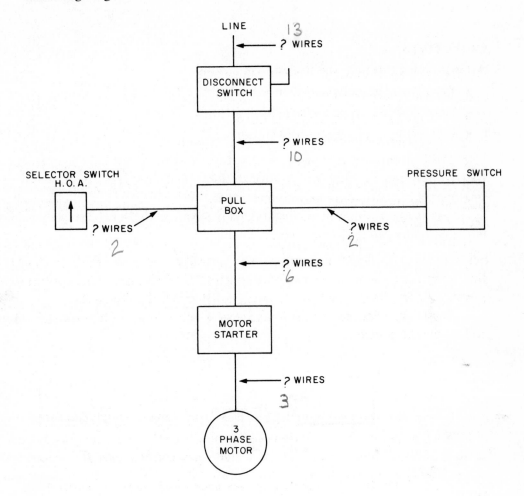

Unit 19 Multiple Pushbutton Stations

OBJECTIVES

After studying this unit, the student will be able to

- Read and interpret diagrams using multiple pushbutton stations.
- Draw diagrams using multiple pushbutton stations.
- Connect multiple pushbutton stations.

The conventional three-wire, pushbutton control circuit may be extended by the addition of one or more pushbutton control stations. The motor may be started or stopped from a number of separate stations by connecting the start buttons in parallel and the stop buttons in series. The operation of each station is the same as that of the single pushbutton control in the basic three-wire circuit covered in unit 17.

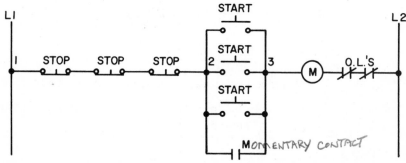

Fig. 19-1 Three-wire control using a momentary contact, multiple pushbutton station.

When a motor must be started and stopped from more than one location, any number of start and stop buttons may be connected. Another possible arrangement is to use only one start-stop station and several stop buttons at different locations to serve as emergency stops.

Multiple pushbutton stations are used to control conveyor motors on large shipping and receiving freight docks.

STUDY/DISCUSSION QUESTIONS

1. Using the line diagram below, determine the number of wires controlling the three-phase motor. The conduit layout plan indicates the sections of conduit for which the quantity of wires is to be determined.

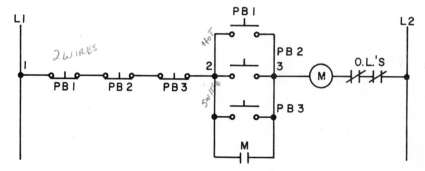

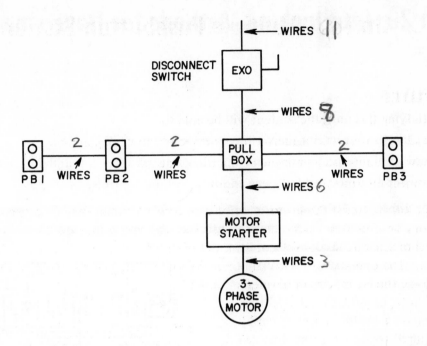

2. Select the minimum size conduit for each run. Use the appropriate locally enforced electrical code — national, state, county, or city.

3. Connect the pushbuttons in parallel in the following diagram with fixed, nonremovable jumpers.

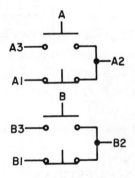

Unit 20 Interlocking Methods for Reversing Control

OBJECTIVES

After studying this unit, the student will be able to

- Explain the purpose of the various interlocking methods.
- Read and interpret wiring and line diagrams of reversing and interlocking controls.
- Wire and troubleshoot reversing and interlocking controls.

The direction of rotation of three-phase motors can be reversed by interchanging any two motor leads to the line. If magnetic control devices are to be used, then the reversal of the motor direction is accomplished by using reversing starters. Reversing starters wired to NEMA standards interchange lines L1 and L3. To do this, two contactors for the starter assembly are required — one for the forward direction and one for the reverse direction. A technique called *interlocking* is used to prevent the contactors from being energized simultaneously or closing together and causing a short circuit. There are three basic methods of interlocking.

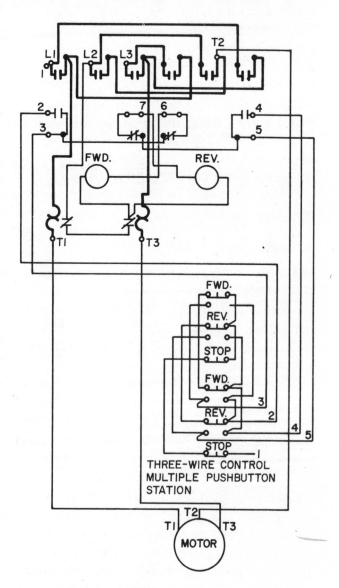

Fig. 20-1 Horizontally mounted reversing starter (left). Wiring diagram of the reversing starter is shown on the right. (Vertical starter arrangements are also available.) The mechanical lever linkage interlock is visible at the front of the starter. The wiring diagram represents the physical location of the power and control wiring and the components of the starter, but not the motor and pushbottons. (Courtesy Square D Co.)

MECHANICAL INTERLOCK

The mechanical interlock shown between the contactors in figure 20-1 is represented in the elementary diagram, figure 20-2, by the broken line between the coils. The broken line indicates that coils F and R cannot close contacts simultaneously because of the interlocking action of the device.

When the forward contactor coil (F) is energized and closed, the mechanical interlock prevents the closing of coil R. Starter F is blocked by coil R in the same manner. The first coil to close moves a lever to a position that prevents the other coil from closing its contacts when it is energized. If the second coil is allowed to remain energized without closing its contacts, the excess current in the coil due to the lack of the proper inductive reactance will dam-

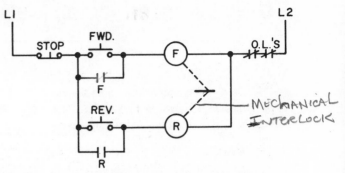

Fig. 20-2 Mechanical interlock between the coils prevents the starter from closing all contacts simultaneously. Only one contact at a time can close.

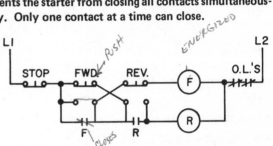

Fig. 20-3 Double-circuit pushbuttons are used for pushbutton interlocking.

age the coil. Note in the wiring diagram of figure 20-1 that the stop button must be pushed before the motor can be reversed.

A mechanical interlock is installed on the majority of reversing starters in addition to the use of one or both of the other interlocking methods.

PUSHBUTTON INTERLOCK

Pushbutton interlocking is an electrical method of preventing both starter coils from being energized simultaneously.

When the forward button in figure 20-3 is pressed, coil F is energized and the normally open (N.O.) contact F closes to hold in the forward contactor. Because the normally closed (N.C.) contacts are used in the forward and reverse pushbutton units, it is unnecessary to press the stop button before changing the direction of rotation. If the reverse button is pressed while the motor is running in the forward direction, the forward control circuit is deenergized and the reverse contactor is energized and held closed.

Repeated reversals of the direction of motor rotation are not recommended since it may cause the overload relays and starting fuses to overheat and thus disconnect the motor from the circuit. Damage to the driven machine also may result.

AUXILIARY CONTACT INTERLOCK

Another method of interlock is obtained through the use of normally closed auxiliary contacts on the forward and reverse contactors of a reversing starter.

When the motor is running forward, an N.C. contact (F) on the forward contactor opens and prevents the reverse contactor from being energized and closing. The same operation occurs if the motor is running in reverse.

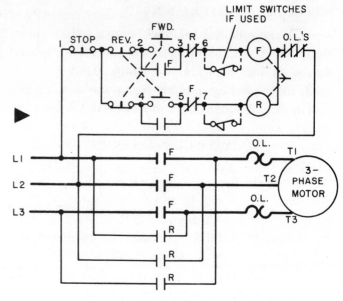

Fig. 20-4 Elementary diagram of the reversing starter shown in figure 20-1. The mechanical pushbutton and auxiliary contact interlocks are indicated.

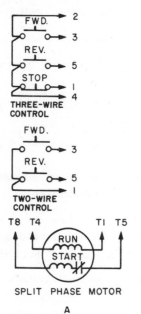

A

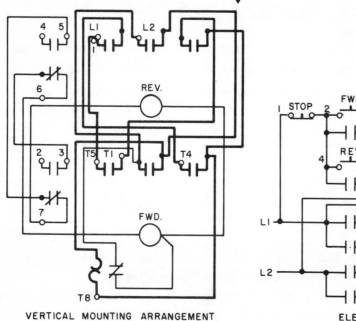

VERTICAL MOUNTING ARRANGEMENT

B

Fig. 20-5 Sizes 0 and 1, three-pole reversing starters used with single-phase, four-wire, split phase induction motors.

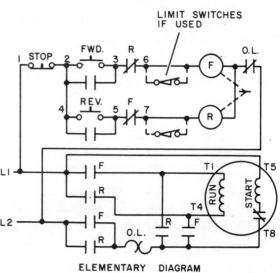

ELEMENTARY DIAGRAM

C

The term interlocking also is used generally when referring to motor controllers and control stations that are interconnected to provide control of production operations. To reverse the direction of rotation of single-phase motors *either* the starting *or* running winding motor leads are interchanged, but not both. Figure 20-5B is a wiring diagram of a single-phase vertical starter; figure 20-5A completes the wiring diagram for the total installation; and figure 20-5C illustrates in line diagram form the same connections.

STUDY/DISCUSSION QUESTIONS

1. How is a change in the direction of rotation of a three-phase motor accomplished? *INTERCHANE ANY TWO MOTOR LEADS TO LINE*

2. What is the purpose of interlocking? *PREVENT CONTACTORS FROM BEING ENERGIZED AT SAME TIME & CAUSING A SHORT CKT.*

3. What will happen if both start buttons are pushed in a control with pushbutton interlocking? Why? *TO PREVENT BOTH STARTER COILS FROM BEING ENERGIZED THE N.O. CONTACTS WILL NOT CLOSE TOGETHER* *NOTHING*

4. How is auxiliary contact interlocking obtained on a reversing starter? *N.C. CONTACT ON FORWARD CONTACTOR OPENS PREVENTS REVERSE CONTACTOR FROM BEING ENERGIZED AND CLOSING*

5. When the forward coil is energized, in what position is the forward interlock (F)? *OPEN*

6. If a mechanical interlock is the only means of interlocking used, describe the operation that must be followed to reverse the direction of rotation of the motor while running. *The STOP BUTTON MUST BE PUSHED BEFORE The MOTOR CAN BE REVERSED.*

7. If pilot lights are to indicate the direction of rotation of a motor, where should the devices be connected so as not to add any contacts?

8. What is the sequence of the operations if limit switches are used in figure 20-4?

9. What will happen in figure 20-4 if limit switches are installed and the jumpers from terminals 6 and 7 to the coils are not removed?

10. In place of the pushbuttons in figure 20-2, draw a selector switch for forward-reverse-stop control. Show the target table for this selector switch.

Unit 21 Sequence Control

OBJECTIVES

After studying this unit, the student will be able to

- State the purpose of starting motors in sequence.
- Read and interpret sequence control diagrams.
- Make the proper connections to operate motors in sequence.
- Troubleshoot sequence motor control circuits.

The method by which starters are connected so that one cannot be started until the other is energized is called sequence control. This type of control is necessary where the auxiliary equipment associated with a machine, such as high pressure lubricating and hydraulic pumps, must be operating before the machine itself can be operated safely.

The proper pushbutton station connections for sequence control are shown in figure 21-1.

Note in the figure that the control circuit of the second starter is wired through the maintaining contacts of the first starter. As a result, the second starter is prevented from starting until after the first starter is energized. If standard starters are used, the connection wire (X) must be removed from one of the starters.

If sequence control is to be provided for a series of motors, the control circuits of the additional starters can be connected in the manner shown in figure 21-1. That is, M3 will be connected to M2 in the same step arrangement by which M2 is connected to M1.

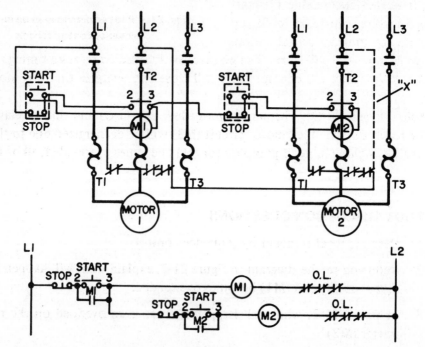

Fig. 21-1 Standard starters wired for sequence control. (Courtesy Allen-Bradley Co.)

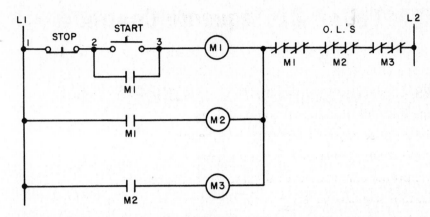

Fig. 21-2 Auxiliary contacts (or interlocks) used for automatic sequence control. Contact (M1) energizes coil (M2); contact (M2) energizes coil (M3).

The stop button (M1) or an overload on motor 1 will stop both motors. The stop button (M2) or an overload on motor 2 will stop only the second motor.

AUTOMATIC SEQUENCE CONTROL

A series of motors can be started automatically with only one start-stop control station as shown in figures 21-2 and 21-3. When the lube oil pump, (M1) in figure 21-3, is started by pressing the start button, the pressure must be built up enough to close the pressure switch before

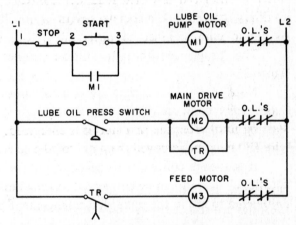

Fig. 21-3 Pilot devices used in an automatic sequence control scheme.

the main drive motor (M2) will start. The pressure switch also energizes a timing relay (TR). After a preset time delay, the contact (TR) will close and energize the feed motor starter coil (M3).

If the main drive motor (M2) becomes overloaded, the starter and timing relay (TR) will open with the result that the feed motor circuit (M3) will be deenergized due to the opening of the contact (TR). If the lube oil pump motor (M1) becomes overloaded, all of the motors will stop.

STUDY/DISCUSSION QUESTIONS

1. Describe what is meant by sequence control.

2. Referring to the diagram in figure 21-2, explain what will happen if the motor on the coil (M1) becomes overloaded.

3. In figure 21-2, what will happen if there is an overload on the motor starter (M2)?

4. What is the sequence of operation in figure 21-2?

Unit 22 Time-Delay, Low-Voltage Release Relay

OBJECTIVES

After studying this unit, the student will be able to

- State the purpose of a time-delay, low-voltage release relay.
- Describe the construction and operation of a time-delay, low-voltage release relay.
- Make the proper connections to insert this type of relay in a circuit containing a motor starter and a pushbutton station.
- Read and interpret control diagrams containing a low-voltage release relay.

The possibility of damage to machinery or injury to an operator makes it desirable to provide some means of preventing motors from restarting when the power suddenly resumes after a prolonged voltage failure. One method of preventing motor restart is known as low-voltage protection or three-wire control. However, since momentary line voltage fluctuations occur in many localities, or there are instances where the voltage actually fails for only a few seconds, three-wire control is not practical. Remember that with three-wire control, each time such a failure occurs the motors must be manually restarted. This is a time consuming procedure and, in some instances, can cause damage to material in process.

One solution to this problem is to use a time-delay, low-voltage release device. When this device is used with a magnetic starter and a momentary contact pushbutton station, the motor is automatically reconnected to the power lines after a voltage failure of short duration. If the voltage failure exceeds the time-delay setting of the low-voltage release device, or if the stop button is pressed, the motor will not restart automatically but must be restarted by pressing the start button. As a result, the use of a time-delay, low-voltage release device provides the safety of three-wire control without the inconvenience of a loss of production time where momentary voltage failures are common.

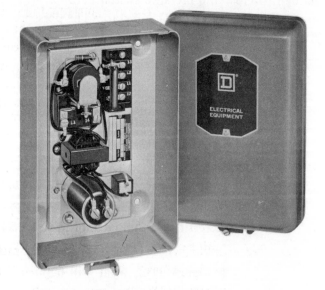

Fig. 22-1 Time-delay, low-voltage release relay. (Courtesy Square D Co.)

CONSTRUCTION AND OPERATION

The time-delay, low-voltage release device consists of a single-pole, normally open control relay, an electrolytic capacitor, a rectifier, two resistors, and a control transformer, figure 22-2. Resistor (R1) is connected in parallel with the coil of the control relay (CR1) to provide a time delay of approximately two seconds. Removal of this resistor provides a time delay of approximately four seconds.

When used with a magnetic starter and a momentary contact, start-stop pushbutton station, the coil of the low-voltage release relay is connected to the secondary of the transformer through the rectifier and a resistor. Current flows through the relay coil at all times, but the value of resistor (R2) is such that the current flowing through the coil is too low to allow relay pickup.

When the start button is pressed, resistor (R2) is cut out, full voltage is applied to the relay coil, and the relay is energized. (When the start button is released, the resistor is again in series with the relay coil. The relay, however, is still energized since less current is required to hold the relay in once it is picked up.) A relay contact, wired in series with the operating coil of the magnetic starter, closes and allows the operation of the magnetic starter.

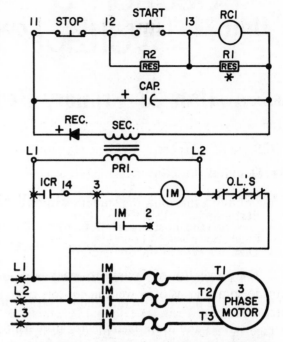

✗ INDICATES TERMINALS ON MAGNETIC STARTER.
✳ TO INCREASE TIME DELAY REMOVE RESISTOR RI.
o INDICATES TERMINALS ON TIME-DELAY, LOW-VOLTAGE RELEASE.

Fig. 22-2 Elementary diagram of low-voltage release relay in figure 22-1.

If the line voltage is reduced or fails completely, the electrolytic capacitor (which is charged through the rectifier) discharges through the control relay coil to keep the coil energized. The time required for the capacitor to discharge is a function of the resistance in the circuit and the capacitance of the device.

If the line voltage returns to approximately 85 percent of its normal value before the capacitor is discharged, the magnetic starter automatically recloses and restarts the motor. If the voltage does not return to normal before the capacitator is discharged, the control relay opens. As a result, the pushbutton must be pressed when power is resumed to restart the motor.

At any time, pushing the stop button causes the control relay to be deenergized and the starter circuit to be opened immediately.

STUDY/DISCUSSION QUESTIONS

1. In the event of line voltage fluctuations or failure, how is the starter maintained in a closed position?

2. How is this principle applied to a dc starter?

3. Under what condition will the motor restart automatically after a voltage failure?

4. Explain in detail the purpose of a time-delay, low-voltage release relay assembly installed in a motor control circuit.

Section 5 AC Reduced Voltage Starters

Unit 23 Primary Resistor-Type Starters

OBJECTIVES

After studying this unit, the student will be able to

- Describe the most important factors to consider when selecting motor starting equipment.
- Describe typical starting methods.
- State why reduced current starting is important.
- Describe the construction and operation of primary resistor starters.
- Interpret and draw diagrams for primary resistor starters.
- Connect squirrel cage motors to primary resistor starters.
- Troubleshoot electrical problems on primary resistor starters.

There are several factors to consider when selecting the starting equipment for a squirrel cage motor installation. The following factors are the most important.

- The torque and starting requirements of the load.
- The motor characteristics that will fit these requirements.
- The power source and the effect the motor starting current will have on the line voltage.
- The effect of the motor starting torque on the driven load.

The simplicity, ruggedness, and reliability of squirrel cage motors have made them the standard choice for alternating-current, all-purpose, constant speed motor applications. Because there are several types of motors available, various kinds of starting methods and control equipment are obtainable also.

TYPICAL STARTING METHODS

Among the most common methods of starting polyphase squirrel cage motors are:

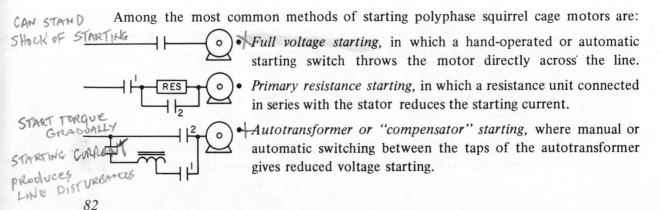

CAN STAND SHOCK OF STARTING

- *Full voltage starting*, in which a hand-operated or automatic starting switch throws the motor directly across the line.

- *Primary resistance starting*, in which a resistance unit connected in series with the stator reduces the starting current.

START TORQUE GRADUALLY

STARTING CURRENT PRODUCES LINE DISTURBANCES

- *Autotransformer or "compensator" starting*, where manual or automatic switching between the taps of the autotransformer gives reduced voltage starting.

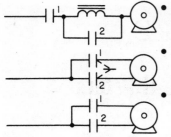

- *Impedance starting,* in which reactors are used in series with the motor.
- *Star-delta starting,* in which the stator of the motor is star connected for starting and delta connected for running.
- *Part winding starting,* in which the stator windings of the motor are made up of two or more circuits; the individual circuits are connected to the line in series for starting and in parallel for normal operation.

Of these methods, the two most fundamental ways of starting squirrel cage motors are full voltage starting and reduced voltage starting. Full voltage starting can be used where the load can stand the shock of starting and objectionable line disturbances are not created. Reduced voltage starting may be required if the starting torque must be applied gradually or if the starting current produces objectionable line disturbances.

Torque Requirements Vary

Different loads have different torque requirements, particularly when considering the starting torque required and the rate of acceleration most desirable for the load. In general, a number of motors will satisfy the load requirements of an installation under normal running conditions. However, it may be more difficult to select a motor that will perform satisfactorily during both starting and running. Often, it is necessary to decide which is the more important consideration for a particular application. For example, a motor may be selected to give the best starting performance but there may be a sacrifice in the running efficiency; or, to obtain a high running efficiency, it may be necessary to select a motor with a high current inrush. In these cases, the selection of the proper starter to overcome the objectionable features is very important.

The machine to which the motor is connected may be started at no load, normal load, or overload conditions. Many industrial applications require that the machine be started when it is not loaded so that the only torque required is that necessary to overcome the inertia of the machine. Other applications may require that the motor be started while the machine it is driving is subjected to the same load it handles during normal running; in this instance, the starting requirements include the ability to overcome both the normal load and the starting inertia.

The problem of providing the proper starting equipment is not always one of merely selecting a starter that satisfies the horsepower rating of the motor; nor is it one of connecting the motor directly across the line. The motor itself must be selected to meet the torque requirements of the industrial application. Actually, the starting equipment often is selected to provide adequate control of the torque after the motor is selected.

THE MOTOR

The Revolving Field

The squirrel cage motor consists of a fixed frame or stator which carries the stator windings and a rotating member which is called the *rotor*. The rotor is constructed of steel laminations mounted rigidly on the motor shaft. The rotor winding consists of many copper or aluminum bars fitted into slots in the rotor. The bars are connected at each end by a continuous ring. The assembly of the rotor bars and end rings resembles a squirrel cage and thus gives the motor its name.

For a three-phase motor, the stator frame has three windings; two windings are used for a two-phase motor. The stator for a squirrel cage motor never has fewer than two poles. The stator windings are connected to the power source. When a 60-hertz current flows in the stator winding, a magnetic field is produced. This field circles the rotor at a speed equal to the number of revolutions per minute (3,600 rpm) divided by the number of pairs of stator poles. Therefore, on 60 hertz, a motor having one pair of two poles will run at 3,600 rpm; a four-pole motor (two pairs of two poles each) will run at 1,800 rpm (3,600/2). The revolving stator field induces current in the short-circuited rotor bars; the current induced has its largest value when the rotor is at a standstill and then decreases as the motor comes up to speed. By changing the resistance and reactance of the rotor, the characteristics of the motor can be changed. For any one rotor design, however, these characteristics are fixed. There are no external connections to the rotor.

No-Load Rotor Speed

. The induced current in the rotor gives rise to magnetic forces which cause the rotor to turn in the direction of rotation of the stator field. The motor accelerates until the necessary speed is reached to overcome windage and friction losses. This speed is referred to as the *no-load speed.* The motor never reaches synchronous speed since a current will not be induced in the rotor under these conditions and thus the motor will not produce a torque.

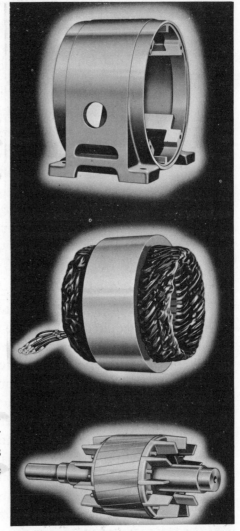

Fig. 23-1 Squirrel cage induction motor frame, stator, and rotor. Note the cooling blades on rotor. (Courtesy U.S. Electrical Motors.)

Speed Under Load

As the rotor slows down under load, the speed adjusts itself to the point where the forces exerted by the magnetic field on the rotor are sufficient to overcome the torque required by the load. The difference between the speed of the magnetic field and the speed of the rotor is known as *slip.*

The slip necessary to carry the full load depends on the motor characteristics. In general, the following situations are true: the higher the inrush current, the lower the slip at which the motor can carry full load, and the higher the efficiency; the lower the value of in-rush current, the higher the slip at which the motor can carry full load, and the lower the efficiency.

✳ An increase in line voltage causes a decrease in the slip, while a decrease in line voltage causes an increase in the slip. In either case, sufficient current is induced in the rotor to carry the load. A decrease in the line voltage has the effect of increasing the heating of the motor. An increase in the line voltage decreases the heating; in other words, the motor can carry a larger load. The slip at rated load may vary from 3 percent to 20 percent for different types of motors.

Locked Rotor Currents

The locked rotor current and the resulting torque are the factors which determine whether the motor can be connected across the line, or whether the current must be reduced to obtain the required performance. Locked rotor currents for different motor types vary from 2 1/2 to 10 times their full load current. Some motors, however, have even higher inrush currents.

Controlling Torque

The most common method of starting a polyphase, squirrel cage induction motor is to connect the motor directly to the plant distribution system at full voltage, using either a manual or magnetic starter. From the standpoint of the motor itself, this is a perfectly acceptable practice; as a matter of fact, it is probably the most desirable method of starting this type of motor.

The motor itself is rarely the final consideration when selecting a starting method. Overload protective devices have reached such a degree of reliability that the motor is given every opportunity to make a safe start. The application of a reduced voltage to a motor in an attempt to prevent overheating during acceleration is generally wasted effort. The accelerating time will increase, and correctly-sized overload elements will still trip.

The need for starting methods other than full voltage starting is largely the result of factors external to the motor. However, it should be noted that once it has been determined that some other starting method is required, the type of motor, its electrical and mechanical characteristics, and its thermal capacity are important considerations in the final selection of the starting method.

Reduced Voltage, Reduced Current, Reduced Torque

In general, all starting methods which deviate from standard line voltage starting are assigned to the category of *reduced voltage* methods. Actually, not all of these starting schemes reduce the voltage at the motor terminals. Even reduced voltage starters reduce the voltage only to achieve either the reduction of line current or the reduction of starting torque. The reduction of line current is the most commonly desired result.

However, the student should note one important point. Failure to remember this point results in many misapplications of starting methods. When the voltage is reduced to start a motor, the current is also reduced, and so is the torque that the machine can deliver. Regardless of the desired result, either reduced current or reduced torque, remember that the other will always follow.

If this fact is kept in mind, it is apparent that a motor which will not start a load on full voltage can not start that same load at reduced voltage or reduced current conditions. The attempt to use a reduced voltage or current scheme for loads that are troublesome to accelerate will fail, since the very process of reducing the voltage and current will further reduce the available starting torque.

The Need for Reduced Current Starting

The most common function of the reduced voltage starting devices presently available is to reduce or in some other way modify the starting current of an induction motor. This function is also expressed as confining the rate of change of the starting current to certain prescribed limits, or predetermining the current-time picture that the motor presents to the supply network.

A current-time picture for an entire area is maintained and regulated by the public power utility serving the area. The power company attempts to maintain a reasonably constant voltage at the points of supply so that lamp flicker will not be noticeable. The success of the power company in this attempt depends on the generating capacity to the area, transformer and line loading conditions and adequacies, and the automatic voltage regulating equipment in use. Voltage regulation also depends on the sudden demands imposed on the supply facilities by residential, commercial, and industrial customers. Transient overloading of the power supply may be caused by sudden high surges of reactive current from large motors on starting, pulsations in current taken by electrical machinery driving reciprocating compressors and similar machinery, the impulse demands of industrial X-ray equipment, and the variable power factor of electric furnaces. All of these demands are capable of producing voltage fluctuations.

Consequently, each of these particularly difficult loads is regulated in some way by the power company which not only dictates the use of some form of reduced voltage, reduced current method, but also helps the customer to determine the best specific method.

Although power company rules and regulations will vary between individual companies and areas served, the following list gives the most commonly applied regulations. This list does not attempt to give all of the possible restrictions on energy usage. Furthermore, an installation may be governed by just one of these restrictions, or two or more rules may be combined for a particular installation.

1. A maximum number of starting amperes, either per horsepower or per motor, is stated.
2. A maximum horsepower for line starting is given. Above this value a limit in percent of full load current is set.
3. A maximum current in amperes for a particular feeder size is given. It is up to the user to determine whether or not the motor will conform to the power company requirements.
4. A maximum rate of change of line current taken by the motor is stated, such as 200 amperes per half-second.

It should be apparent that it is very important for the electrician to understand the behavior of an induction motor during the startup and acceleration periods to enable him to select the proper starting method to conform to local power company regulations. While several starting methods may appear to be appropriate, a careful examination of the specific application will usually indicate the one best method for motor starting.

The Induction Motor At Start

Figure 23-2 illustrates the behavior of the current taken by an induction motor at various speeds. First of all, note that the starting current is high compared to the running current. In addition, the starting current remains fairly constant at this high value as the motor speed increases and then drops sharply as the motor approaches its full rated speed. Since the motor heating rate is a function of I^2, this rate is high during acceleration. Also, the motor can be considered to be in the locked condition during most of the acceleration period.

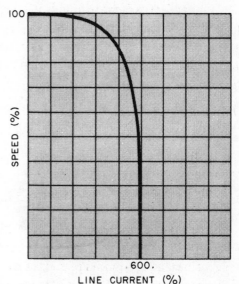

Fig. 23-2 Induction motor current at various speeds.

PRIMARY RESISTOR-TYPE STARTERS

A simple method of starting a motor at reduced voltage is used in primary resistor-type starters. This method has a resistor connected in series in the line to the motor. Thus, the motor starting speed and current are reduced. The resistor can be disconnected when the motor reaches a certain speed and the motor is then connected to run on full line voltage. The introduction and removal of resistance in the motor starting circuit may be accomplished manually or automatically.

Primary resistor starters are used to start squirrel cage motors where a limited torque is necessary to avoid damage to driven machinery. These starters are also used with limited current inrush to prevent excessive power line disturbances.

It is desirable to limit the starting current when the power system does not have the capacity for full voltage starting and when full voltage starting may cause serious line disturbances, such as in lighting circuits, electronic circuits, the simultaneous starting of many motors, or in cases where the motor is distant from the incoming power supply. For these situations, reduced voltage starters may be recommended for motors with ratings as small as five horsepower.

Reduced voltage starting must be used for driving machinery which must not be subjected to sudden high starting torque and the shock of sudden acceleration. Typical applications include those where belt drives may slip or where large gears, fan blades, or couplings may be damaged by sudden starts.

Automatic primary resistor starters use either one or more than one step of acceleration, depending upon the size of the motor being controlled. These starters provide smooth acceleration without the line current surges normally experienced when switching autotransformer types of reduced voltage starters.

Special starters are required for very high inertia loads with long acceleration periods or where power companies require that current surges be limited to specific increments at stated intervals.

Figures 23-3, 23-4, and 23-5 on pages 88 and 89 illustrate an automatic, primary resistor-type, reduced voltage starter. Figure 23-5 shows the starter connected to a three-phase, squirrel cage induction motor.

When the start button is pressed, a complete circuit is established from L1 through the stop button, start button, coil (M), and the overload relay contacts to L2. When coil (M) is energized, the main power contacts (M) and the control circuit maintaining contact (M) are closed. The motor is energized through the overload heaters and the starting resistors. Because the resistors are connected in series with the motor terminals, a voltage drop occurs in the resistors and the motor starts on reduced voltage.

As the motor accelerates, the voltage drop across the resistors decreases gradually because of a reduction in the starting current (E = IR). At the same time, the motor terminal voltage increases.

After a predetermined acceleration time, delay contact (M) closes the circuit to contactor coil (S). Coil (S), in turn, closes contacts (S), the resistances are shunted out, and the motor is connected across the full line voltage.

Fig. 23-3 Primary resistor starter for reduced voltage starting. Resistors are mounted in rear vented compartment. (Courtesy Square D Co.)

Note that the stop button controls coil (M) directly. When the main power contacts (M) open, coil (S) drops out. After coil (M) is energized, a pneumatic timing unit attached to the starter unit (M) retards the closing of time contact (M). This scheme uses starter (M) for a dual purpose and eliminates one coil, a timing relay.

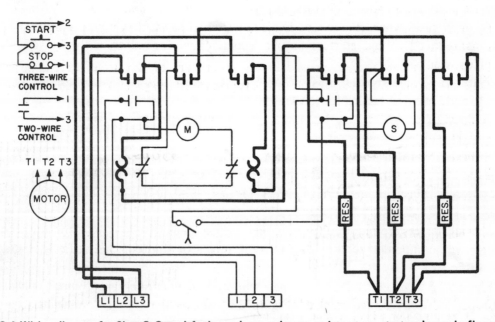

Fig. 23-4 Wiring diagram for Sizes 2, 3, and 4, three-phase, primary resistor-type starter shown in figure 23-3.

Quizz
Know This

MAINTAIN
CONTACT
CONTROL

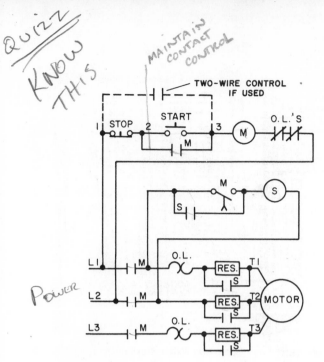

Fig. 23-5 Line diagram of figure 23-4 for primary resistor starter.

Power

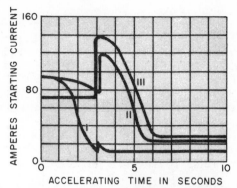

Fig. 23-6 The curves illustrate how a primary resistance starter reduces the starting current of a 10 h.p., 230-volt motor under three different load conditions. Curve I is at a light load and Curve II is at a heavy load. In either case, the running switch closes after three seconds. With a heavy load, an increase in the starting time reduces the second current inrush; the motor will reach a higher speed before it is connected to full line voltage. In Curve III the motor cannot start on the current allowed through the resistance; it comes up to speed only after connection to full line voltage.

For maximum operating efficiency, pushbuttons or other pilot devices are usually mounted on the driven machinery within easy reach of the operator. The starter is located near the motor to keep the heavy power circuit wiring as short as possible. Only two or three small connecting wires are necessary between the starter and pilot device. A motor can be operated from any one of several remote locations if a number of pushbuttons or pilot switches are used with one magnetic starter.

Ac primary resistor starters are available for use on single-phase and three-phase reversing operations.

The timers or timing relays used in the circuit are of the preset time type, such as pneumatic or dashpot timers. Compensating-type current relays are also used.

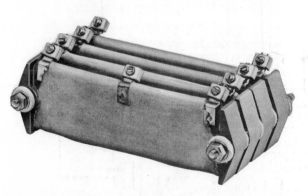

Fig. 23-7 Resistors used in primary resistor-type starters. Ribbon resistor (right) is used with larger starters. (Courtesy Square D Co.)

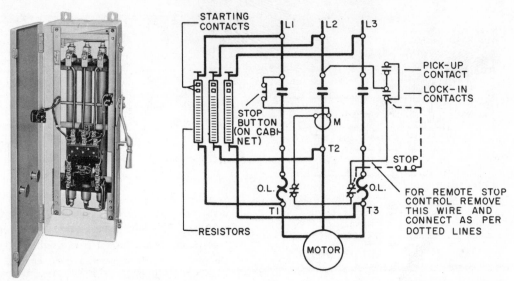

Fig. 23-8 Manual resistance starter with "carbon pile" resistors. (Courtesy Allen-Bradley Co.)

The resistor banks used for starting small motors consist of resistance wire wound around porcelain bases and imbedded in refractory cement. The resistor shown at the right of figure 23-7 has a ribbon-type construction and is used on larger motor starters. This type of resistor consists of an alloy ribbon formed in a zig-zag shape. The formed ribbon is supported between porcelain blocks which have recesses for each bend of the ribbon. The ribbons are not stressed and do not have a reduced cross section, since they are bent on the flat and not on the edge. Any number of units can be combined vertically or horizontally, in series or in parallel.

Some starters have graphite compression disk resistors and are used for starting polyphase squirrel cage motors, figure 23-8.

Features of a primary resistance-type starter include simple construction, low initial cost, low maintenance, smooth acceleration in operation, continuous connection of the motor to the line during the starting period, and a high power factor. These starters should not be used for starting very heavy loads because of their low starting torque.

STUDY/DISCUSSION QUESTIONS

1. What is the purpose of inserting resistance in the stator circuit during starting? *1.Reduces The STARTING CURRENT SOFT START Keep TORQUE DOWN*
   Primary Secondary

2. Why is the power not interrupted when the motor makes the transition from start to run? *Because of The PRIMARY RESISTANCE STARTER CONTINUOS CONNECTION TO THE MOTOR TO THE LINE (SHORT OUT RESISTANCE)*

3. If the starter is to function properly, and a timing relay is used, where must the coil of the relay be connected? *PARALLEL WITH M COIL*

4. How may additional steps of acceleration be added? *3 STEP STARTER ADD RELAY & CONTACTORS TO CUT OUT STEPS*

5. What is meant by the low or poor starting economy of a primary resistor starter? *LOW TORQUE*

6. List five commonly used starting methods.
   1.FULL VOLTAGE STARTING
   2 AUTOTRANSFORMER
   3. PRIMARY - RESISTANCE
   4. STAR-DELTA
   5. PART-WINDING

Unit 24 Autotransformer Starters

REDUCED VOLTAGE STARTER

OBJECTIVES

After studying this unit, the student will be able to

- Describe the construction and operation of autotransformer starters.
- Draw and interpret diagrams for autotransformer starters.
- Connect squirrel cage motors to autotransformer starters.
- Define what is meant by open transition and closed transition starting.
- Troubleshoot electrical problems on autotransformer starters.

Autotransformer reduced voltage starters are similar to primary resistor starters in that they are used primarily with ac squirrel cage motors to limit the inrush current or to lessen the strain on driven machinery.

To reduce the voltage across the motor terminals during the accelerating period, an autotransformer-type starter generally has two autotransformers connected in open delta. During the reduced voltage starting period, the motor is connected to taps on the autotransformer. Due to the lower starting voltage, the motor draws less current and develops less torque than if it were connected to line voltage, figure 24-3.

An adjustable time relay controls the transfer from the reduced voltage condition to full voltage. A current-sensitive relay may be used to control the transfer to obtain current-limiting acceleration.

To understand the operation of the autotransformer starter more clearly, refer to the line diagram in figure 24-2, page 92. When the start button is closed momentarily, the timing relay (TR) is energized and maintains the circuit across the start button with the normally open instantaneous contact (TR). Starting coil (S) is energized from terminal four through the "time-delay in opening" contact (TR) and through the normally closed interlock (R). The running starter can

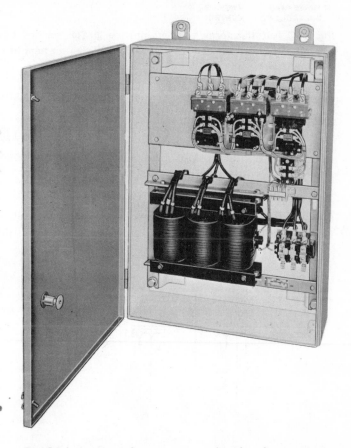

Fig. 24-1 Autotransformer-type, reduced voltage starter. Field connecting terminals for power and control are visible at bottom. (Courtesy Allen-Bradley Co.)

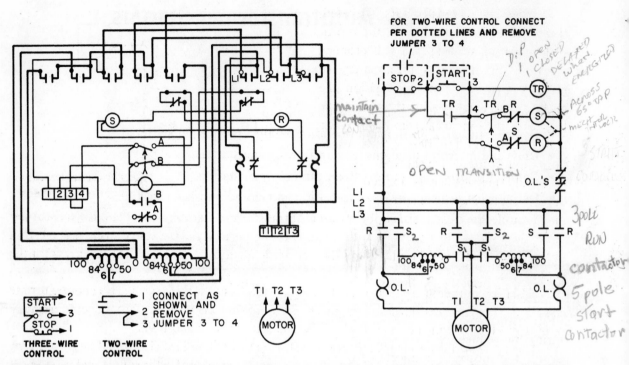

KNOW FOR QUIZZ

Fig. 24-2 Autotransformer starters provide greater starting torque per ampere drawn from the line than any other type of reduced voltage starter. Wiring diagram of figure 24-1 is shown on left, while the line diagram is shown on the right.

not be closed at this point because the normally closed interlock (S) is open and the mechanical interlock is operating.

After a preset timing period (TR), the normally closed contacts open and the normally open (TR) contacts close. When the coil (S) is deenergized, N.C. interlock (S) closes and energizes the running starter (R).

Figure 24-3A shows the power circuit for starting the motor with two autotransformers, and figure 24-3B shows the circuit for starting a motor with three autotransformers. The contact switching arrangement for a typical power circuit is shown in figure 24-2.

KNOW These!

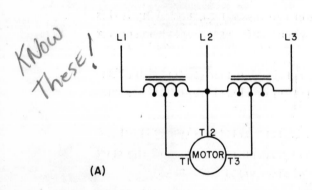

(A)

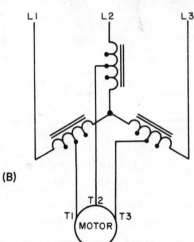

(B)

Fig. 24-3 Power circuit connections showing two and three autotransformers used for reduced voltage starting.

[handwritten: Know This]

[handwritten: Starting contactors]

[handwritten: 2 pole - open]
[handwritten: 3 pole - closed]

When two transformers are used, there will be an imbalance in the motor voltage during starting that will produce a torque variation of approximately 10 percent. In the running position the motor is connected directly across the line and the autotransformers are disconnected from the line. As a result, only three contacts are shown.

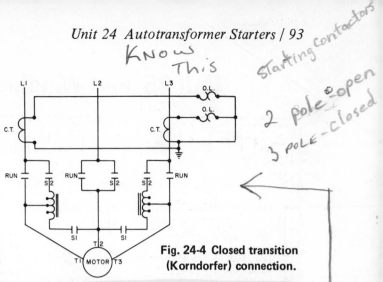

Fig. 24-4 Closed transition (Korndorfer) connection.

Full line voltage is applied to the outside terminals of the autotransformer on starting. Reduced voltage for starting the motor is obtained from the autotransformer taps. Generally, there are a number of taps brought out from the autotransformers so that different voltages may be obtained. The current taken by a motor varies directly with the applied voltage.

[handwritten left margin: Five pole / 3 pole]

Starting compensators (autotransformer starters) using a five-pole starting contactor are classified as open transition starters due to the fact that the motor is disconnected momentarily from the line during the transfer from the start to run conditions.

Closed transition connections are usually found on standard size 6 and larger starters. For the closed transition starter, the starting contactors consist of a three-pole and a two-pole contactor operating independently of each other. During the transfer from start to run, the two-pole contactor is open and the three-pole contactor remains closed. The motor continues to accelerate with the autotransformer serving as a reactor. With this type of starter, the motor is not disconnected from the line during the transfer period. Thus, there is less line disturbance and a smoother acceleration.

The (CT) designations in figure 24-4 indicate current transformers. These transformers are used on large motor starters to step down the current so that a conventional overload relay size may be used. Magnetic overload relays are used on large reduced voltage starters also.

STUDY/DISCUSSION QUESTIONS

1. Why is it desirable to remove the autotransformers from the line when the motor reaches its rated speed? *[handwritten: WILL NOT DRAW MAGNETIC CURRENT]*

2. What is meant by an "open transition" from start to run? Why is this condition objectionable at times when used with large horsepower motors? *[handwritten: MOTOR IS DISCONNECTED MOMENTARILY FROM LINE - LINE DISTURBANCE]* *[handwritten: OPEN ONE BEFORE OTHER CONTACT D.C. START BEFORE RUN]*

3. Which of the following applies to an autotransformer starter with a five-pole starting contactor: open transition or closed transition? Locate in the diagrams of figure 24-2. *[handwritten: OPEN TRANSITION 5 5 POLES]*

4. How are reduced voltages obtained from autotransformer starters? *[handwritten: FROM THE AUTOTRANSFORMER TAPS]*

5. If the motor is running and the stop button is pressed, and the start button then is pressed immediately, what happens? *[handwritten: MOTOR CONTINUES AFTER SLIGHT HESITATION - START ALL OVER]*

6. What is a disadvantage of starting with autotransformer coils rather than with resistors? *[handwritten: HIGH EXCITING CURRENT ON TRANSFORMER]*

7. What is one advantage of using an autotransformer starter? *[handwritten: MOTOR IS NOT DISCONNECTED FROM THE LINE DURING THE TRANSFER PERIOD FROM START TO RUN. PROVIDE GREATER STARTING TORQUE PER AMP FROM THE LINE. PER DRAWN]*

Unit 25 Part Winding Motor Starters

OBJECTIVES

After studying this unit, the student will be able to

- Describe the construction and operation of a part winding motor starter.
- Draw and interpret diagrams for part winding motor starters.
- List the advantages and disadvantages of two-step part winding starters.
- Connect motors to part winding starters.
- Troubleshoot part winding motor starters.

Part winding motors are similar in construction to standard squirrel cage motors. However, part winding motors have two identical windings that may be connected to the power supply in sequence to produce reduced starting current and reduced starting torque. Since only half of the windings are connected to the supply lines at startup, the method is described as *part winding*. Many, but not all, dual-voltage, 220/440-volt motors are suitable for part winding starting at 220 volts.

Part winding starters are designed to be used with squirrel cage motors having two separate and parallel stator windings. In the part winding motor, these windings may be Y-connected or delta-connected, depending upon the motor design. Part winding starters are not suitable for use with delta-wound, dual-voltage motors.

Part winding motors are used to drive centrifugal loads such as fans, blowers, or centrifugal pumps, and for other loads where a reduced starting torque is necessary. This type of motor is also used where the full voltage starting current will produce objectionable voltage drops in the distribution feeders, or where power company restrictions require a reduced starting current.

Air-conditioning systems provide many applications for this type of starting because of the increased capacity built into these systems, coupled with the necessity of limiting both the current and torque upon starting.

Fig. 25-1 Size 4, two-step, part winding motor starter
(Courtesy Square D Co.)

TWO-STEP STARTING

In a two-step part winding starter, figure 25-2, pressing the start button energizes the start contactor (S). As a result, half of the motor windings are connected to the line and a timing relay (TR) is energized. After a time delay of

94

no more than five seconds, the timer contacts (TR) close to energize the "run" contactor. This contactor connects the second half of the motor windings to the line in parallel with the first half of the windings. Note that the control circuit is maintained with an instantaneous N.O. contact (TR) operated by the timing relay (TR) and the starting contact (S).

Pressing the stop button or tripping any overload relay disconnects both windings from the line.

When starting on one winding, the motor draws approximately two-thirds of the normal locked rotor current, and develops approximately one-half of the normal locked rotor torque.

A two-step part winding starter has certain obvious advantages: it is less expensive than most other starting methods as it requires no voltage reducing elements such as transformers, resistors, or reactors; it uses only half-size contactors; and it provides closed transition starting.

However, the two-step part winding starter also has disadvantages: the fixed starting torque is poor, and the starter is almost always an incremental start device.

Part winding starters do not possess the flexibility of application that other, more expensive, types of reduced voltage starters have.

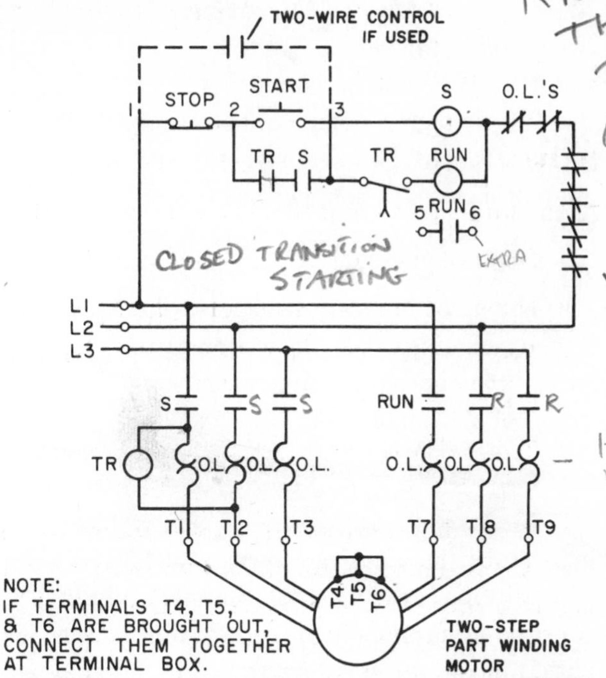

THREE-STEP STARTING

The thermal capacity of the motor limits the length of acceleration on the first winding to approximately five seconds. In many instances, the motor will not begin to accelerate until the second winding is connected.

Three-step starting is similar to two-step starting with the exception that when the first contactor closes, the first winding is connected to the line through a resistor in each phase.

Fig. 25-2 Line diagram of a two-step, part winding starter. When dual-voltage, Y-connected motors are used on the lower voltage only, terminals T4, T5, and T6 are connected to form the center of the second Y.

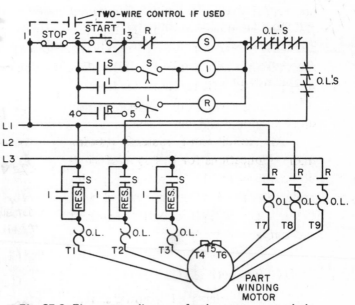

Fig. 25-3 Elementary diagram of a three-step, part winding motor starter.

After a time delay of approximately two seconds, this resistor is shorted out and the first winding is connected to the full voltage. After another time delay of about two seconds, the run contactor closes to connect both windings to the line voltage. The resistors are designed to provide approximately 50 percent of the line voltage to the starting winding. Thus, the motor starts with three approximately equal increments of starting current.

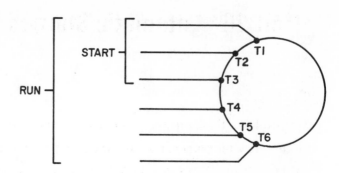

Fig. 25-4 Part winding, single-voltage, star or delta motor terminal markings.

OVERLOAD PROTECTION

Motor overload protection is provided by six overload relays, three in each motor winding. Since the windings are identical, each winding should have the same full load current. In addition, each phase of the start winding will carry a current identical to that of the corresponding phase in the run winding. Thus, if three overload relays are connected in three phases in the run winding and three other relays are connected in three different phases in the start winding, then the effect of full three-phase protection is obtained.

When selecting heater elements for the overload relays, the electrician should remember that each of the relays carries only half of the motor current. The elements, therefore, must be selected on the basis of one-half of the full load current of the motor.

The current requirements in part winding starters permit the use of smaller contactors with smaller overload elements. As a result, the starting fuses must be proportionally smaller to protect these elements. Thus, dual element fuses are a necessity. Local electrical codes and ordinances should always be consulted when in doubt regarding installation and protection problems.

STUDY/DISCUSSION QUESTIONS

1. How can a 220/440-volt motor be used as a part winding motor?

2. Which voltage may be applied to a 220/440-volt motor if used with a part winding starter? Why?

3. If the nameplate of a dual voltage motor connected to a part winding starter reads 15/30 amperes, on what amperage value is the selection of the overload heating elements made?

4. If a timing relay fails to cut in the run starter, what percentage of the full load current can the motor safely carry?

5. When starting on one winding, a motor draws approximately what percentage of the normal locked rotor current?

6. How is a third step gained in part winding starting?

7. Who determines the compliance of an installation to applicable electrical codes with regard to adequate overload or starting and short circuit protection?

Unit 26 Automatic Starters for Star-Delta Motors

OBJECTIVES

After studying this unit, the student will be able to

- Identify terminal markings for a star-delta motor and motor starter.
- Describe the purpose and function of star-delta starting.
- Troubleshoot star-delta motor starters.
- Connect star-delta motors and starters.

Star-delta motors are similar in construction to standard squirrel cage motors, with the exception that both ends of each of the three windings are brought out to the terminals. If a starter is used which has the required number of properly wired contacts, the motor can be started in star and run in delta.

The first requirement of this scheme is that the motor be wound in such a manner that it will run with its stator windings connected in delta. The leads of all the windings must be brought out to the motor terminals for their proper connection in the field.

The primary applications of star-delta motors are for driving centrifugal loads such as fans, blowers, pumps, or centrifuges, and for situations where a reduced starting torque is necessary. Star-delta motors also may be used where a reduced starting current is required.

The synchronous speed of a star-delta, squirrel cage induction motor depends upon the number of poles of the motor and the supply line frequency. Since both of these values are constant, the motor will run at approximately the same speed for either

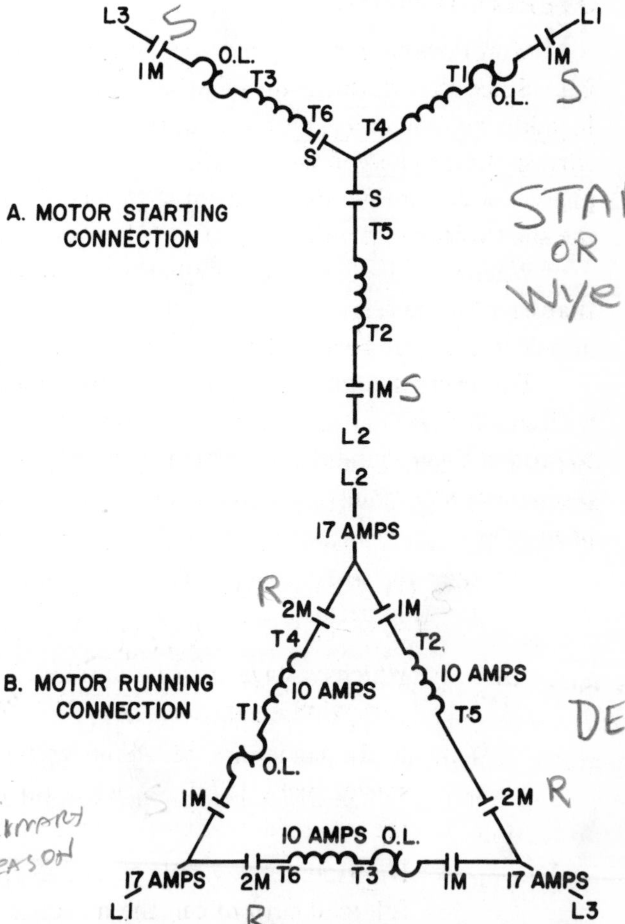

Fig. 26-1 Elementary diagrams of motor power circuits of figure 26-2. Controller connects motor in wye on start and in delta for run. Note that the overload relays are connected in the motor winding circuit, not in the line. Note also that the line current is higher than the phase winding current in the diagram for the delta connection (B). Winding current is the same as the line current in diagram A.

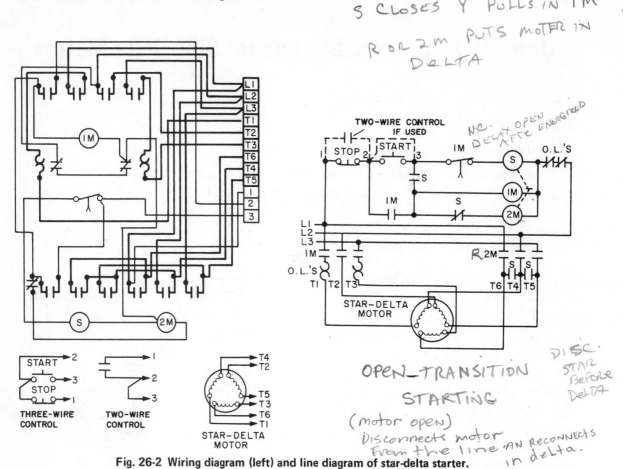

Handwritten annotations:

S CLOSES Y PULLS IN 1M
R OR 2M PUTS MOTER IN
DELTA

N.C. OPEN
DELAY OPEN
AFTER ENERGIZED

OPEN-TRANSITION
STARTING
(motor open)
Disconnects motor
from the line, AN RECONNECTS
in delta.

DISC.
STAR
BEFORE
DELTA

Fig. 26-2 Wiring diagram (left) and line diagram of star-delta starter.

the star or delta connection. The inrush and line current are less when the motor is connected in star than when it is connected in delta, and the winding current is less than the line current when the motor is connected in delta. That is, the inrush and line current in star are one-third the values of these quantities in delta; the winding current in star is 1.73 times the winding current in delta.

Three overload relays are furnished on star-delta starters. These relays are wired so that they carry the motor winding current, figure 26-1, page 97. Thus, the relay units must be selected on the basis of the winding current, not the delta-connected full load current. If the motor nameplate indicates only the delta-connected full load current, divide this value by 1.73 to obtain the winding current which is used as the basis for selecting the motor winding protection.

OPERATION

Open transition starting for a star-delta starter is shown in figure 26-2. As indicated in the line diagram on the right, the automatic transfer from star to delta is accomplished by a pneumatic timer, operated by the movement of the armature of one of the contactors. Operating the pushbutton station start button energizes contactor (S) whose main contacts connect three of the motor leads together (T4, T5, and T6) to form a star. Normally open, control contact (S) of the same contactor energizes another contactor (1M). Since the pneumatic timer is attached to contactor (1M) the motor is connected to the line in star and the timing period is started. When the timing period is complete, the first contactor (S) is deenergized, with the result that normally closed interlock (S) is closed, and contactor (2M) is energized to connect the motor in delta. The motor then runs in the delta-connected configuration. This start-run scheme is called open transition because there is a moment in which the motor circuit is open between the opening of the power contacts (S) and the closing of the contacts (2M).

One advantage of this starting method is that it does not require accessory voltage reducing equipment. While a star-delta starter has the disadvantage of open circuit transition, it does give a larger starting torque per line ampere than a part winding starter.

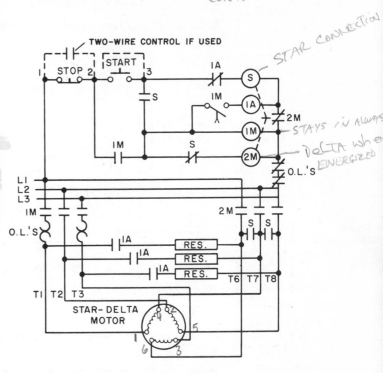

Fig. 26-3 Elementary diagram of Sizes 1, 2, 3, 4, and 5 star delta-starters with closed transition starting.

CLOSED TRANSITION STARTING

Figure 26-3 shows a modification of figure 26-2. In figure 26-3, resistors are used to maintain continuity to the motor to avoid the difficulties associated with the open circuit form of transition between start and run.

With closed transition starting, the transfer from the star to delta connections is made without disconnecting the motor from the line. When the transfer from star to delta is made in open transition starting, the starter momentarily disconnects the motor and then reconnects it in delta. While an open transition is satisfactory in many cases, some installations may require closed transition starting to prevent power line disturbances. Closed transition starting is achieved by adding a three-pole contactor and three resistors to the starter circuit, connected as shown in the closed transition schematic diagram, figure 26-3. The contactor is energized only during the transition from star to delta. It keeps the motor connected to the power source through the resistors during the transition period. The incremental current surge, which results from the transition is thus reduced. The balance of the operating sequence of the closed transition starter is similar to that of the open transition star-delta motor starter.

The motor starting requirements may be so involved, restrictions so stringent, and needs so conflicting that it may not be possible to achieve the desired results with a single method of reduced current starting. It may be necessary to resort to a combination of starting methods before satisfactory performance is realized. Special installations require the design of a starting system that is tailor-made to fit the particular conditions.

STUDY/DISCUSSION QUESTIONS

1. Indicate the correct terminal markings for a star-delta motor on the diagram below.

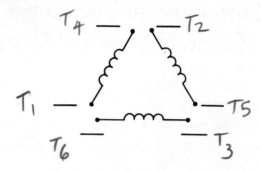

2. What is the principal reason for using star-delta motors?

3. In figure 26-3, which contactor closes the transition?

4. The closed transition contactor is energized only on transfer from star to delta. How is this accomplished?

5. If a delta-connected, six-lead motor nameplate reads "Full Load Current 170 Amperes," upon what current rating should the overload relay setting, or the selection of the heater elements be based?

Section 6 Three-Phase, Multispeed Controllers
Unit 27 Controllers for Two-Speed, Two-Winding (Separate Winding) Motors

OBJECTIVES

After studying this unit, the student will be able to

- Identify terminal markings for two-speed, separate winding motors and starters.
- Describe the purpose and function of two-speed, two-winding motor starters.
- Describe how different speeds for ac squirrel cage motors are obtained.
- Connect two-speed, two-winding controllers and motors.
- Troubleshoot motors and controllers.

THE MOTOR

The squirrel cage induction motor is the most widely used industrial motor because it is simple and less expensive than most other motors. Essentially, it is a constant speed motor. Some models, however, are made to operate at several fixed speeds. To do this, the motor manufacturer changes the number of poles for which the stator is wound. The operating principle on which the squirrel cage motor is based calls for a rotor revolving within a stator due to the action of a rotating magnetic field. The speed of the revolving field is determined by the frequency of the alternating current supplied by the alternator and the number of poles on the stator. The formula defining speed in rpm is,

$$\text{Speed in rpm} = \frac{60 \times \text{Frequency}}{\text{Pairs of Poles}}$$

The abbreviation rpm (revolutions per minute) refers to the synchronous speed of the motor (rotor speed equals rotating magnetic field speed). This synchronous speed actually is not achieved due to slip (lag of the rotor behind the rotating stator field). The phrase *pairs of poles* is used to indicate that motor poles are always in pairs; a motor never has an odd number of poles.

As an example of the use of the above formula, determine the synchronous speed of a two-pole motor supplied by 60-hertz electrical energy.

$$\frac{60 \times 60}{1} = 3,600 \text{ rpm}$$

The motor will run just below the synchronous speed due to slip. The design of the motor will determine the percentage of slip. This value is not the same for all motors.

Fig. 27-1 Magnetic controller for two-speed, two-winding (separate winding) motor. (Courtesy Square D Co.)

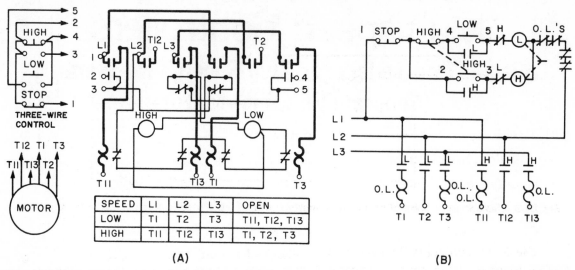

Fig. 27-2 Wiring diagram (A) and line diagram (B) for Sizes 2, 3, 4, and 5, two-speed, two-winding (separate winding), three-phase motor starters.

CONTROLLERS

Line voltage, ac multispeed starters are designed to provide control for squirrel cage motors operating on two, three, or four different constant speeds, depending upon their construction. The use of an automatic starter and a control station results in greater operating efficiency and offers protection, both to the motor and the machine, against improper sequencing or speed change occurring too rapidly. Protection against motor overload in each speed is necessary.

Motors with separate windings have a different winding for each speed required. While the construction of this type of motor is slightly more complicated and thus more expensive, the controller is relatively simple.

Assuming the frequency is constant, two winding motors will operate on each winding at the speed for which they are wound, figure 27-1.

In the control arrangement in figure 27-2B, note that the motor can be started in either speed, but the stop button must be pressed to transfer from the high speed to the low speed. The starter shown in the figure is provided with a mechanical interlock between the L and H starters; auxiliary contact interlocks are also provided.

Figure 27-3 illustrates pushbutton interlocking. The figure also indicates that a transfer can be made to either speed without touching the stop button. However, rapid and continued speed transfers may activate the overload relays. To prevent this, the control scheme can be equipped with time-delay relays that will provide a time lag between the speed changes.

When the motor is operating, the inactive winding must be open circuited to prevent circulating current arising from the transformer action between the idle winding and the energized winding.

Figure 27-4 illustrates the function of an accelerating relay circuit. If the high speed is desired, the high-speed button is depressed, and control relays (CR) are energized. The circuit is maintained through the normally open contact under the high pushbutton located from contact 2 to contact 3, resulting in the opening of normally closed contact (CR).

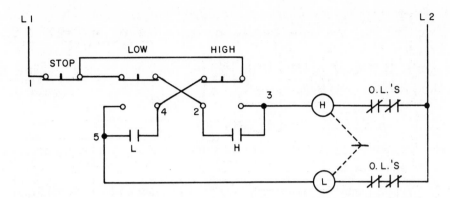

Fig. 27-3 Elementary diagram of pushbutton interlocking and transfer to either speed without stopping.

The N. O. contact (CR), (under the low pushbutton) from contact 2 to contact 4, energizes the low-speed starter coil (L) through N.C. contacts (TR and H). While the motor is starting on low speed, contact (CR) closes N. O. auxiliary contact (L) to energize the timing relay (TR). The circuit is maintained through instantaneous N. O. contact (TR) and contact 2. When the N. C., delay-in-opening contact (TR) opens after a predetermined time lag, the N. O. delay-in-closing contact (TR) closes. As a result, the starter coil (L) opens and the N. C. interlock (L) in the coil circuit (H) is closed. Once the coil (H) is energized, the motor is running on high speed.

To start the motor and maintain it at a low speed, press the low pushbutton. The motor is started in low speed

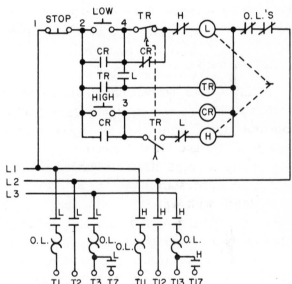

Fig. 27-4 Elementary diagram of starter with an accelerating relay for a two-speed, two-winding, delta-connected, three-phase motor.

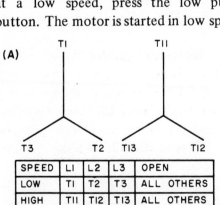

(A)				
SPEED	L1	L2	L3	OPEN
LOW	T1	T2	T3	ALL OTHERS
HIGH	T11	T12	T13	ALL OTHERS

(B)				
SPEED	L1	L2	L3	OPEN
LOW	T1	T2	T3, T7	ALL OTHERS
HIGH	T11	T12	T13, T17	ALL OTHERS

Fig. 27-5 Connections for three-phase, two-speed, two-winding (separate winding) motors (A) star (wye) and (B) delta.

and the coil (TR) is energized and maintained through the instantaneous N. O. contact (TR). The normally closed delay-in-opening contact now opens, but coil (L) is maintained through N. C. contact (CR).

To change to high-speed operation, the high button can be pressed because the N. O. delay-in-opening contact (TR) is closed also.

STUDY/ DISCUSSION QUESTIONS

1. Excluding slip, at what speed (rpm) will an eight-pole, 60-hertz, 240-volt motor run?

2. What is meant by the description "separate winding, two-speed motor"?

3. What will happen if an operator makes too many rapid transitions from low to high speed and back on a two-speed motor?

4. Why are there four motor contacts for high and low speeds in figure 27-4?

5. In figure 27-4, what does the broken line between both time-delay contacts (TR) signify?

6. When the "high" button is depressed in figure 27-4, what prevents the high-speed starter from closing immediately?

7. Which coils are energized immediately when the high pushbutton is closed in figure 27-4?

8. At what speed can a four-pole, 240-volt, 5-h.p., 50-hertz motor operate?

9. What is meant by the "compelling action" produced by an accelerating relay?

10. Is it necessary to press the stop button when changing from a high speed to a low speed in figure 27-4?

Unit 28 Two-Speed, One-Winding (Consequent Pole) Motor Controller

OBJECTIVES

After studying this unit, the student will be able to

- Identify terminal markings for two-speed, one-winding (consequent pole) motors and controllers.
- Describe the purpose and function of two-speed, one-winding motor starters and motors.
- Connect and troubleshoot two-speed, one-winding controllers and motors.
- Connect a two-speed starter with reversing controls.

Certain applications require the use of a motor which has a winding arranged so that the number of poles can be changed by reversing some of the currents. If the number of poles is doubled, the speed of the motor is cut approximately in half.

The number of poles can be cut in half by changing the polarity of alternate pairs of poles, figure 28-1. The polarity of half the poles can be changed by reversing the current in half the coils.

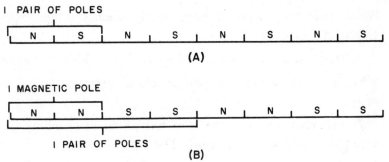

Fig. 28-1 (A) Eight poles for low speed; (B) Four poles for high speed.

If a stator field is laid flat as in figure 28-1, it can be seen that the established stator field must move the rotor twice as far in A as in B in the same amount of time; therefore, it must travel faster. The fewer the number of poles established in the stator, the greater is the speed in rpm of the rotor.

A three-phase motor can be wound so that six leads are brought out. By making suitable connections with these leads, the windings can be connected in series delta or parallel wye. If the winding is such that the series delta connection gives the high speed and the parallel wye connection gives the low speed, then the horsepower

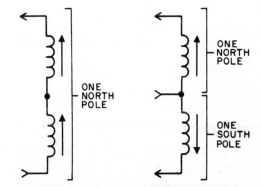

Fig. 28-2 The number of poles is doubled by reversing current through half a phase. Two speeds are obtained by producing twice as many consequent poles for low-speed operation as for high speed.

rating is the same at both speeds. If the winding is such that the series delta connection gives the low speed and the parallel wye connection gives the high speed, then the torque rating is the same at both speeds.

Consequent pole motors have a single winding for two speeds. Extra taps can be brought from the winding to permit reconnection for a different number of stator poles. The speed range is limited to a 1:2 ratio, such as 600-1200 rpm or 900-1800 rpm.

Two-speed, consequent pole motors have one reconnectable winding; three-speed, consequent pole motors have two windings, one of which is reconnectable; four-speed consequent pole motors have two reconnectable windings.

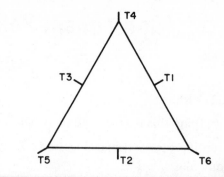

SPEED	LI	L2	L3	OPEN	TOGETHER
LOW	TI	T2	T3	——————	T4, T5, T6
HIGH	T6	T4	T5	ALL OTHERS	——————

Fig. 28-3 Connection table for three-phase, two-speed, one-winding, constant horsepower motor.

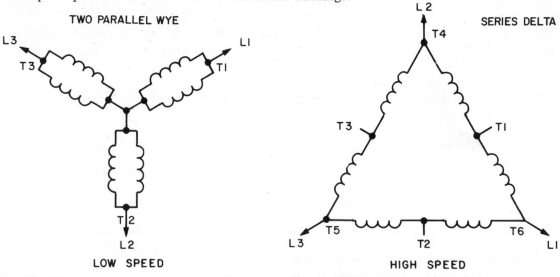

Fig. 28-4 Three-phase, two-speed, one-winding, constant h.p. motor connections made by motor controller.

Referring to the motor connection table in figure 28-3, note that for low speed operation, T1 is connected to L1, T2 to L2, T3 to L3, and T4, T5, and T6 are connected together. For high speed operation, T6 is connected to L1, T4 to L2, T5 to L3, and all other motor leads are open.

Figure 28-5 illustrates a Size 1, selective multispeed starter connected for operation with a reconnectable, constant horsepower motor. The control is a three-element, fast-slow-stop station connected for starting in either the fast or slow speed. In addition, the speed can be changed from fast to slow or slow to fast without pressing the stop button first. If equipment considerations make it desirable to stop the motor before changing speeds, this feature can be added to the control circuit by making connection D, shown in figure 28-5A and B by the dashed lines.

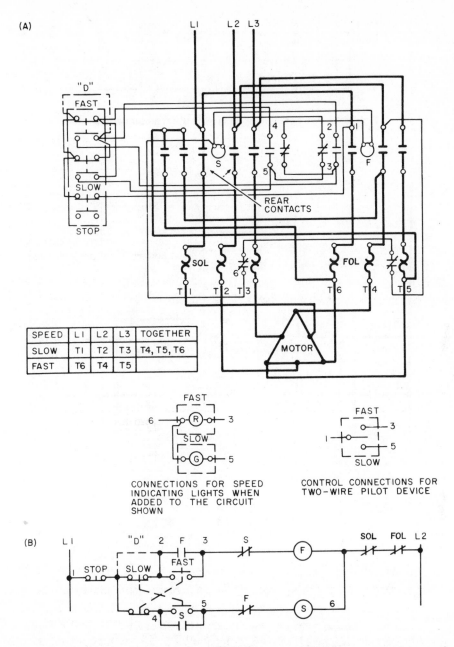

SPEED	L1	L2	L3	TOGETHER
SLOW	T1	T2	T3	T4, T5, T6
FAST	T6	T4	T5	

Fig. 28-5 Wiring diagram (A) and line diagram (B) of an ac, full voltage, two-speed magnetic starter for single-winding (reconnectable pole) motors.

Connections for the addition of indicating lights or a two-wire pilot device instead of the control shown are also given.

A compelling-type control scheme is shown in figure 28-6, page 108. This controller is wired so that the operator must start the motor at the slow speed. This controller cannot be switched to the fast speed until after the motor is running.

When the slow button is pressed, the slow-speed starter (S) and the control relay (FR) are energized. Once the motor is running, pressing the fast button causes the slow-speed

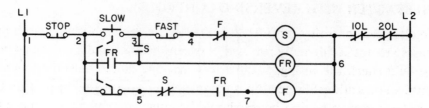

Fig. 28-6 Two-speed control circuit using a compelling relay.

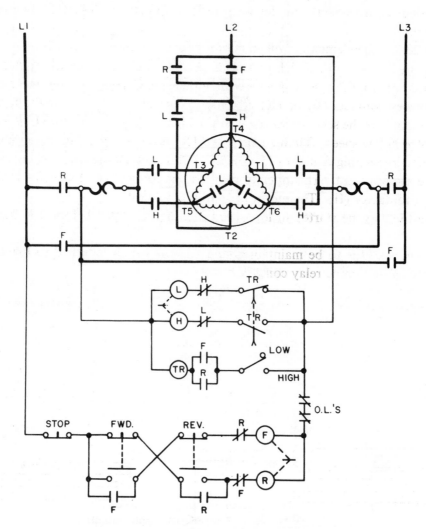

Fig. 28-7 Elementary diagram of a two-speed, reversing controller.

starter to drop out while the high-speed starter is picked up through the N. C. interlock contacts of the slow-speed starter and the N. O. contacts of the control relay (FR).

If the fast button is pressed in an attempt to start the motor, nothing will happen because the N. O. contacts of the control relay (FR) will prevent the high-speed starter from energizing. When the fast button is pressed, it breaks a circuit but does not make any. This is another form of sequence starting.

TWO-SPEED STARTER WITH REVERSING CONTROLS

When multispeed controllers are installed, the electrician should check carefully to insure that the phases are not accidentally reversed between the high- and low-speed windings. The electrician should check the wiring, or the driven machine should remain disconnected from the motor until an operational inspection is completed. The upper oil inspection plugs (pressure plugs) in large gear reduction boxes should be removed. Failure to remove these plugs may create broken casings if the motor is reversed accidentally.

Machines can be damaged if the direction of rotation is changed from that for which they are designed. In general, the correct rotational direction is indicated by arrows on the driven machine.

Figure 28-7 is the elementary diagram of a two-speed, reversing controller.

The desired speed, either high or low, is selected with a two-position selector switch. The direction of rotation is selected with either the forward or reverse pushbuttons. When the power contacts (F) or (R) are closed, current is supplied for the high-low controls. Assuming that the selector switch is in the high position, contactor (L) is energized to start the motor in low speed. Timing relay coil (TR) also is energized at this moment. When the N.C. delay-in-opening contact (TR) opens after a preset time delay, N.O. contact (TR) closes at the same instant. This operation drops out contactor (L), and the N.C. interlock (L) energizes contactor (H). The motor is now at high speed.

The motor may be started in low speed in either direction before it is transferred to high speed.

If the low speed is to be maintained, the selector switch is turned to low to open the circuit supplying the timing relay coil.

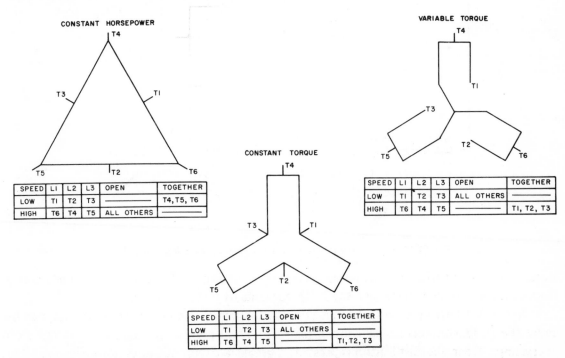

Fig. 28-8 Typical motor connection arrangement for three-phase, two-speed, one-winding motors. These connections conform to NEMA and ASA standards. (All possible arrangements are not shown.)

STUDY/DISCUSSION QUESTIONS

1. The rotating magnetic field of a two-pole motor travels 360 electrical degrees. How many mechanical degrees does the rotor travel?

2. How may two speeds be obtained from a one-winding motor?

3. Referring to figure 28-3, what will happen if terminals T5 and T6 are interchanged on high speed?

4. At what speed (high or low) is the motor operating when it is connected in series delta? Parallel wye?

5. In figure 28-6, the motor cannot be started in high speed. Why?

Unit 29 Four-Speed, Two-Winding (Consequent Pole) Motor Controller

OBJECTIVES

After studying this unit, the student will be able to

- Identify terminal markings for four-speed, two-winding (consequent pole) motors and controllers.
- Describe the purpose and function of four-speed, two-winding motor starters and motors.
- Describe the purpose and function of a compelling relay, an accelerating relay, and a decelerating relay.
- Connect and troubleshoot four-speed, two-winding controllers and motors.

Four-speed, consequent pole motors have two reconnectable windings and two speeds for each winding.

OPERATIONAL SEQUENCE

Standard starters generally are connected so that the operator can start the motor from rest at any speed or change from a lower to a higher speed. The stop button, however, must be pressed before a change is made from a higher to a lower speed. This arrangement is necessary to protect the motor and driven machinery from the excessive line current and shock resulting when a motor running at a high speed is reconnected to run at a lower speed.

Multispeed starters are provided with mechanical and electrical interlocking to avoid the possibility of short circuiting the line or connecting more than one speed winding at the same time.

COMPELLING RELAYS

Standard starters are not equipped with compelling relays which insure that the motor starts at the lowest speed or that acceleration is accomplished in a series of steps. However, when the type of motor used or the characteristics of the load involved makes it necessary to use a particular starting sequence, there are three types of relays available to accomplish the function required.

Form 1 Compelling Relay

When this type of relay is used, the motor must be started at low speed before a higher speed can be selected. The motor can be started only by pressing the low speed pushbutton. If any other pushbutton is pressed, nothing will happen. This arrangement insures that the motor first moves the load only at low speed. The stop button must be pressed before it is possible to change from a higher to a lower speed.

Form 2 Accelerating Relays

When the starter is equipped with a Form 2 accelerating relay, the *ultimate* speed is determined by the button which is pressed. However, the motor is started at low speed and

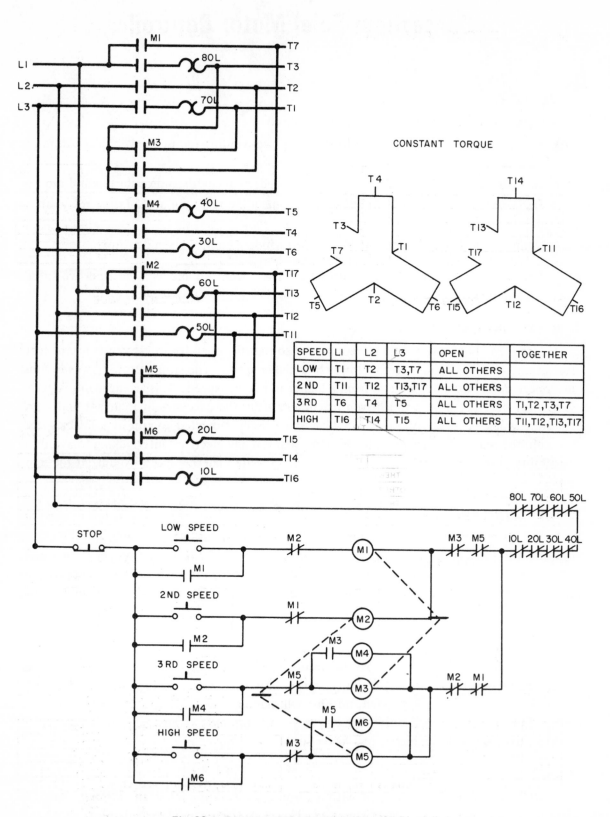

SPEED	LI	L2	L3	OPEN	TOGETHER
LOW	TI	T2	T3,T7	ALL OTHERS	
2ND	TII	T12	T13,TI7	ALL OTHERS	
3RD	T6	T4	T5	ALL OTHERS	TI,T2,T3,T7
HIGH	T16	T14	T15	ALL OTHERS	TII,T12,T13,TI7

Fig. 29-1 Four-speed, two-winding controller.

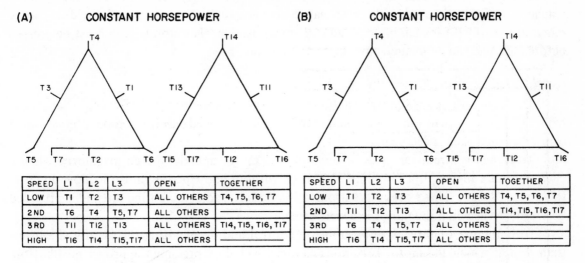

(A) CONSTANT HORSEPOWER

SPEED	LI	L2	L3	OPEN	TOGETHER
LOW	TI	T2	T3	ALL OTHERS	T4, T5, T6, T7
2ND	T6	T4	T5, T7	ALL OTHERS	————
3RD	TII	T12	T13	ALL OTHERS	T14, T15, T16, T17
HIGH	T16	T14	T15, T17	ALL OTHERS	————

(B) CONSTANT HORSEPOWER

SPEED	LI	L2	L3	OPEN	TOGETHER
LOW	TI	T2	T3	ALL OTHERS	T4, T5, T6, T7
2ND	TII	T12	T13	ALL OTHERS	T14, T15, T16, T17
3RD	T6	T4	T5, T7	ALL OTHERS	————
HIGH	T16	T14	T15, T17	ALL OTHERS	————

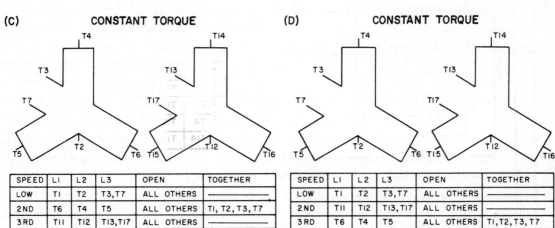

(C) CONSTANT TORQUE

SPEED	LI	L2	L3	OPEN	TOGETHER
LOW	TI	T2	T3, T7	ALL OTHERS	————
2ND	T6	T4	T5	ALL OTHERS	TI, T2, T3, T7
3RD	TII	T12	T13, T17	ALL OTHERS	————
HIGH	T16	T14	T15	ALL OTHERS	TII, T12, T13, T17

(D) CONSTANT TORQUE

SPEED	LI	L2	L3	OPEN	TOGETHER
LOW	TI	T2	T3, T7	ALL OTHERS	————
2ND	TII	T12	T13, T17	ALL OTHERS	————
3RD	T6	T4	T5	ALL OTHERS	TI, T2, T3, T7
HIGH	T16	T14	T15	ALL OTHERS	TII, T12, T13, T17

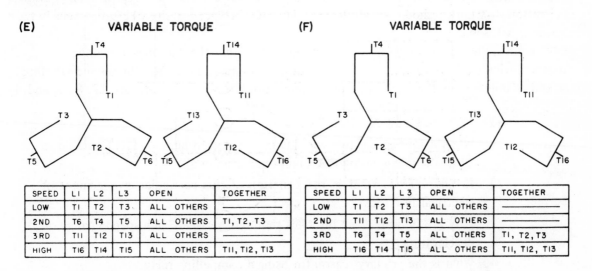

(E) VARIABLE TORQUE

SPEED	LI	L2	L3	OPEN	TOGETHER
LOW	TI	T2	T3	ALL OTHERS	————
2ND	T6	T4	T5	ALL OTHERS	TI, T2, T3
3RD	TII	T12	T13	ALL OTHERS	————
HIGH	T16	T14	T15	ALL OTHERS	TII, T12, T13

(F) VARIABLE TORQUE

SPEED	LI	L2	L3	OPEN	TOGETHER
LOW	TI	T2	T3	ALL OTHERS	————
2ND	TII	T12	T13	ALL OTHERS	————
3RD	T6	T4	T5	ALL OTHERS	TI, T2, T3
HIGH	T16	T14	T15	ALL OTHERS	TII, T12, T13

Fig. 29-2 Typical arrangements for four-speed, three-phase, two-winding motors.

then is accelerated automatically through successive speed steps until the selected speed is reached. Definite time intervals must elapse between each speed change. Individual timing relays are provided for each interval, and all are adjustable. The stop button must be pressed before a change can be made from a higher to a lower speed.

Decelerating Relays

This type of relay is similar in action to a Form 2 accelerating relay, except that it controls a stepped deceleration from a high speed to a lower speed. Definite time intervals must elapse between each speed change.

Tremendous strains are placed on both the motor and the driven machinery when a change from a higher to a lower speed is made without allowing the motor to slow down to the desired speed. These strains are avoided if a definite, adjustable time interval is provided between speeds when the motor is decelerating.

With the use of three-wire pushbutton control and decelerating relays, any speed less than that at which the motor is running can be selected by pressing the proper pushbutton. Once this pushbutton is pressed, it deenergizes the contactor driving the motor at the higher speed and energizes the timing relays. After a preset time period, these relays energize the contactor to drive the motor at the lower speed. As succeedingly lower speeds are selected, there is an increase in the elapsed time between disconnecting the higher speed winding and connecting the lower speed winding.

When two-wire control devices, such as pressure or float switches are used to control motors at various speeds, decelerating relays should always be used. The only exception to this statement is in the event that both the motor manufacturer and the machine manufacturer approve the intended application and agree that decelerating relays can be omitted.

Figure 29-1, page 112, is an elementary drawing of a four-speed controller to be used for a two-winding motor. This is a standard starter arrangement which is not equipped with compelling relays that permit starting only at the lowest speed. This controller has electrical interlocks which prevent an operator from changing to different speeds without pressing the stop button. There should be a definite time interval between each speed change on deceleration; the motor should be allowed to slow to the speed desired before a transfer to a lower speed is made.

Note in the motor connection table in figure 29-1 that the windings are connected alternately during acceleration. If the first winding is 6 and 12 poles and the second winding is 4 and 8 poles, then the successive pole connections for acceleration are 12, 8, 6, and 4.

STUDY/DISCUSSION QUESTIONS

1. When a four-speed, two-winding motor is running, why is it desirable to allow the idle winding to remain open circuited?
2. In figure 29-1, what is the function of this particular arrangement of electrical interlocks?
3. Why are definite time intervals required between each speed change on deceleration?
4. What is the primary reason for using a compelling relay?
5. What is the purpose of an accelerating relay?

6. When is it most important to use decelerating relays for multispeed installations?

7. Why are different motor connections shown for what appears to be the same motor in the various diagrams in figure 29-2, page 113?

8. How many windings are required for three-speed motors?

Section 7 Wound Rotor (Slip Ring) Motor Controllers

Unit 30 Manual Speed Control

OBJECTIVES

After studying this unit, the student will be able to

- Identify terminal markings for wound rotor (slip ring) motors and controllers.
- Describe the purpose and function of manual speed control for wound rotor motors.
- Explain the difference between two-wire and three-wire control for wound rotor motors.
- Connect and troubleshoot wound rotor motors with manual speed controllers.

The wound rotor or slip ring induction motor was the first alternating-current motor that successfully provided speed control characteristics. This type of motor was an important factor in adapting alternating current for industrial power applications. The wound rotor motor has the added features of high starting torque and low starting current. These features give the motor better operating characteristics for applications requiring a large motor or where the motor must start under load. This motor is especially desirable where its size in any given situation is large with respect to the capacity of the transformers or power lines.

The phrase wound rotor actually describes the construction of the rotor, that is, wound with wire. When installed in a motor, three leads are brought out from the rotor winding to solid slip rings. Carbon brushes ride on these rings and carry the rotor winding circuit out of the motor to a controller. The controller varies the resistance in the rotor circuit to control the acceleration and speed of the rotor once it is operating.

Resistance is introduced into the rotor circuit when the motor is started or when it is operating at slow speed. As the external resistance is eliminated by the controller, the motor accelerates.

Motors for school and classroom use are provided with a stator and several interchangeable rotors that can be used with the same stator and bearing end brackets. The selection of rotors available includes the following types: squirrel cage, wound, and synchronous.

A control for a wound rotor motor consists of two separate elements. First, there is a means of connecting the

Fig. 30-1 Rotor of an 800-h.p., wound rotor induction motor. (Courtesy Electric Machinery Mfg. Co.)

primary or stator winding to the power lines, and then there is a mechanism for controlling the resistance in the secondary or rotor circuit. For this reason, wound rotor motor controllers are often called *secondary resistor starters.*

Secondary resistor starters and regulators have a sliding contact (faceplate) design. In this type of regulator, stationary contacts are connected into the resistor circuit. Movable contacts slide over these stationary contacts, from left to right, and cut out of the rotor circuit successive steps of resistance, resulting in an increase in the speed of the motor. The entire resistance is removed from the rotor circuit when the extreme right position of the stationary contacts is reached. At this point, the motor is operating at normal speed. The resistors used in starters are designed for starting duty only; therefore,

Fig. 30-2 Manual, sliding contact (faceplate) secondary resistor regulator for a wound rotor motor has an electrical interlock to control the magnetic primary (stator) contactor. (Courtesy Cutler-Hammer, Inc.)

the operating lever should always be moved to the full ON position. The lever must not be left in any intermediate position. The resistors used in regulators are designed for continuous duty. As a result, the operating lever can be left in any speed position.

Sliding contact wound rotor starters and regulators do not have a magnetic primary contactor; they control the secondary (rotor) circuit of the motor only. A separate magnetic contactor, automatic starter, or circuit breaker is required for the primary circuit. If a primary magnetic contactor is used, electrical interlocks on the movable contact arm will control the operation of the contactor.

When starting wound rotor motors, the operating lever is moved to the first running position to insert the full value of resistance in the secondary circuit, figure 30-3. This action also operates the electrical contact and completes the circuit to the magnetic primary contactor. This contactor connects the primary circuit to the line through the normally open contact (two-wire control). As the operating lever moves to the right, more and more

resistance is cut out of the circuit until, at the extreme right-hand position of the lever, the motor is running at full speed. N. O. or N. C. electrical interlocks control the primary magnetic switch and insure that sufficient resistance is used in the rotor circuit for starting.

The use of two-wire control (normally open contact) for starting provides low voltage release. This means that power to the motor can be

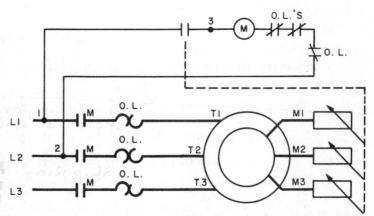

Fig. 30-3 Manual speed regulator interlocked with magnetic starter for control of slip ring motor (two-wire control).

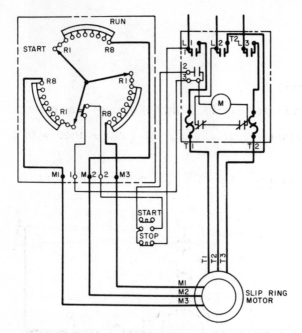

Fig. 30-4 Wiring diagram of manual speed regulator interlocked with magnetic starter for control of slip ring motor (three-wire control).

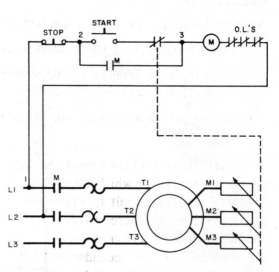

Fig. 30-5 Elementary diagram of figure 30-4.

interrupted if the line voltage drops to a low value or fails completely. When the voltage returns to its normal value, the motor is restarted automatically.

The use of three-wire control for starting (normally closed control contact), figures 30-4 and 30-5, means that low-voltage protection is provided. The motor is disconnected from the line in the event of voltage failure. To restart the motor when the voltage returns to its normal value, it is necessary to follow the normal starting procedure.

Drum controllers may be used in a manner similar to that of faceplate starters. The drum controller, however, is an independent component; that is, it is separate from the resistors. As in the faceplate starter, a starting contact controls a line voltage, across-the-line starter.

STUDY/DISCUSSION QUESTIONS

1. What characteristic of wound rotor motors led to their wide use for industrial applications? *Sucessfully provided speed control under load characteristics.*

2. By what means is the rotor coupled to the stator? *Transformer induction*

3. What other name is given to the wound rotor motor? *Slip Ring*

4. Does increased resistance in the rotor circuit produce low or high speed? *Higher Low speed & torque*

5. What two separate elements are used to control a wound rotor motor? *Connect Primary or stator windings to power lines. Mag. starter speed reg.*

6. What is the difference between a manual faceplate starter and a manual faceplate regulator? *Lower resist start. Resistors used in starters designed for starting duty only, in a regulator the resistors are designed for continous duty. Higher watt resist. (size of resistors)*

Unit 31 Pushbutton Speed Selection

OBJECTIVES

After studying this unit, the student will be able to

- Describe the purpose and function of pushbutton speed selection of wound rotor motors.
- Describe what happens during acceleration and deceleration of a motor used with a pushbutton speed selector control.
- Connect and troubleshoot wound rotor motors and pushbutton speed selector controls.

Magnetic controllers consist of a magnetic starter which connects the primary circuit to the line, and one or more accelerating contactors to cut out gradually the resistance in the secondary circuit. The number of secondary accelerating contactors depends on the rating of the motor. Enough contactors are used to assure a smooth acceleration and to keep the inrush current within practical limits. The accelerating contactors can be wired for operation by pushbutton speed selection, figure 31-1.

When the low-speed pushbutton is pressed, the stator is energized by coil (M) through power contacts (M). The motor starts slowly with the full resistance in the rotor secondary circuit. When contactor (S) is energized by the medium speed pushbutton, contacts (S) shunt out part of the total resistance in the rotor circuit, with the result that the rotor speed increases. The speed will continue to increase if the high-speed button is pressed, because contactor (H) is

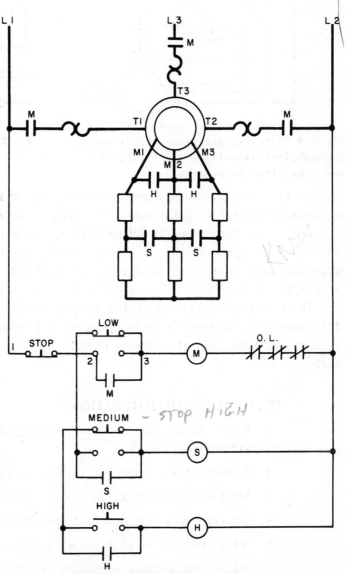

Fig. 31-1 Pushbutton speed selection for a wound rotor motor.

119

energized, contacts (H) are closed, and all resistance in the secondary circuit of the motor is eliminated.

If it is desired to go to medium speed from the high-speed position, the medium push-button is pressed. This action deenergizes contactor (H), inserts part of the total resistance into the rotor circuit, and thus decreases the rotor speed. A transfer to low speed can be made in the same manner by merely pressing the low pushbutton.

This method of pushbutton speed selection is a relatively simple, inexpensive control installation. The resistors should have the capacity (power) to operate at any speed.

A disadvantage of this method is that the motor and the driven machine can be accelerated without allowing sufficient time between steps for the rotor to gain its maximum speed for each step of acceleration. The desired time delay can be achieved by adding compelling relays to the circuit to prevent too rapid an acceleration.

STUDY/DISCUSSION QUESTIONS

1. Can the motor in figure 31-1 be started in medium or high speed? Why?
   NO must go to lower speed first

2. Must the stop button be pressed between speeds on acceleration or deceleration? Why?
   No   High Low

3. When the motor is being accelerated, must each speed pushbutton be closed in succession? Why? *Yes – sequence control*

4. Is it possible to transfer from a high speed to low speed without pressing the medium pushbutton? Why? *Low Drop out   Ω High*
   Yes

5. Do all contactors remain energized when the motor is in high speed?
   Yes

6. Accelerating resistors cannot be used for continuous duty. Why is this true and what may happen if an attempt is made to use them for continuous duty? *Low power Short term*
   Resist. Burn-out

7. Draw a line diagram of a three-speed acceleration control which uses compelling timing relays to prevent too rapid an acceleration.

Unit 32 Automatic Acceleration

OBJECTIVES

After studying this unit, the student will be able to

- State the purpose and advantages of using controllers to provide automatic acceleration of wound rotor motors.
- Identify terminal markings for automatic acceleration controllers used with wound rotor motors.
- Describe the process of automatic acceleration using reversing control.
- Describe the process of automatic acceleration using frequency relays.
- Connect and troubleshoot wound rotor motor and automatic acceleration and reversal controllers.

Secondary resistor starters used for the automatic acceleration of wound rotor motors consist of an across-the-line starter for connecting the primary circuit to the line, and one or more accelerating contactors to shunt out resistance in the secondary circuit as the rotor speed increases. The secondary resistance consists of banks of three uniform wye sections, each of which is to be connected to the slip rings of the motor. The wiring of the accelerating starters and the design of the resistor sections are meant for starting duty only. This type of controller cannot be used for speed regulation. The current inrush on starters with two steps of acceleration is limited by the secondary resistors to a value of approximately 250 percent at the point of the initial acceleration. Resistors on starters with three or more steps of acceleration limit the current inrush to 150 percent at the point of initial acceleration. Resistors for acceleration generally are designed to withstand one 10-second accelerating period in each 80 seconds of elapsed time for a duration of one hour without sustaining damage.

The operation of accelerating contactors is controlled by a timing device that provides timed acceleration. The timing of the steps of acceleration is controlled by adjustable pneumatic accelerating relays. When these relays are properly adjusted, all starting periods are the same regardless of variations in the starting load. This automatic timing feature eliminates the danger of an improper startup procedure by an inexperienced operator.

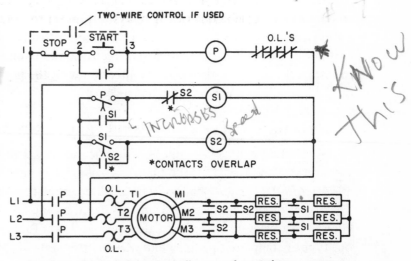

Fig. 32-1 Typical elementary diagram of wound rotor motor starter with three points of acceleration.

121

MAKE DRAWING

The primary circuit (stator) in figure 32-1 is energized with the start button. The motor starts with a full value of resistance in the secondary circuit. Coil (P) actuates the N. O., delay-in-closing contact (P). After a preset timing period, contact (P) closes, energizes contactor (S1), and maintains itself through the maintaining contact (S1). When contacts (S1) in the secondary resistor circuit close, the motor continues to accelerate. After the N. O., delay-in-closing contact (S1) times out, contactor (S2) is energized and closes resistor circuit contacts (S2). The motor then is accelerated to its maximum speed. The N. C. interlock (S2) opens the contactor (S1) circuit. The closing of S2 is assured by staggered, or overlapping, control contacts (S2).

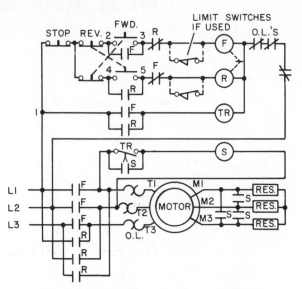

Fig. 32-2 Typical elementary diagram of a starter with two points of acceleration for a reversing wound rotor motor.

AUTOMATIC ACCELERATION WITH REVERSING CONTROL

Automatic acceleration is possible in either direction of rotation with the addition to the circuit of reversing contactors and pushbuttons. The wiring of these devices is shown in figure 32-2.

The motor may be started in either direction of rotation at low speed with the full secondary resistance inserted in the circuit. For either direction of rotation, the timing relay (TR) is energized by the N. O. auxiliary contacts (F or R). Coil (TR) activates the N. O., delay-in-closing contact (TR). Coil (S) is energized when contact (TR) times out and removes all of the resistors from the circuit to achieve maximum motor speed. The primary contactors are interlocked with the pushbuttons, normally closed contacts, and the mechanical devices. If a limit switch is used, connections shown by dashed lines in figure 32-2, the motor will stop when the limit switch is struck and opened. In this situation, it is necessary to restart the motor in the opposite direction with the pushbutton. As a result, lines 1 and 3 on the primary side will be interchanged.

AUTOMATIC ACCELERATION USING FREQUENCY RELAYS

Definite timers or compensated timers may be used to control the acceleration of wound rotor motors. Definite timers, which usually consist of pneumatic or dashpot relays, are set for the highest load current and remain at the same setting regardless of the load. The operation of a compensated timer is based on the applied load; that is, the motor will be allowed to accelerate faster for a light load and slower for a heavy load. The frequency relay is one type of compensating timer and uses the principle of electrical resonance in its operation.

When a 60-hertz ac, wound rotor motor is accelerated, the frequency induced in the secondary circuit decreases from 60 hertz at zero speed to two or three hertz at full speed.

Fig. 32-3 Rotor frequency decreases as the motor approaches full speed.

The voltage between the phases of the secondary decreases in the same proportion from zero speed to full-speed operation. At zero speed, the voltage induced in the rotor is determined by the ratio of the stator and rotor turns. The frequency, however, is the same as that of the line supply. As the rotor accelerates, the magnetic fields induced in it almost match the rotating magnetic field of the stator. As a result, the number of lines of force that the rotor cuts is decreased with a resulting decrease in the frequency and voltage of the rotor. The rotor never becomes fully synchronized with the rotating field because of the slip necessary to achieve the relative motion required for induction and the operation of the rotor. The percentage of slip determines the value of the secondary frequency and voltage. If the slip is five percent, then the secondary frequency and voltage are five percent of normal.

Figure 32-4 illustrates a simplified frequency relay system operated by pushbutton starting. This system has two contactor coils connected in parallel (A and B) and a capacitor connected in series with coil B. A three-step automatic acceleration results from this arrangement. When the motor starts, full voltage is produced across coils A and B, causing normally closed contacts (A and B) to open. The full resistance is connected across the secondary of the motor. As the motor accelerates, the secondary frequency decreases, with the result that coil (B) drops out and contacts (B) close to decrease the resistance in the rotor circuit. The capacitor depends upon the frequency of an alternating current. As the motor continues to accelerate, coil (A) drops out and closes contacts (A). Because the normally closed contacts

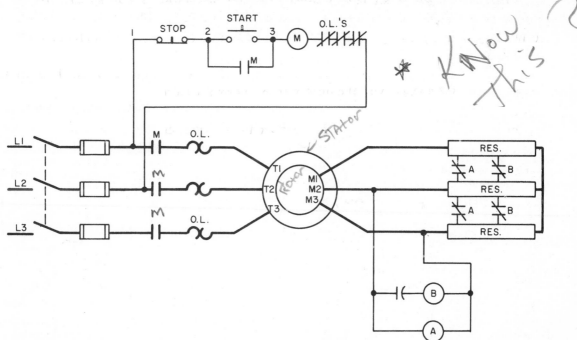

Fig. 32-4 Automatic acceleration of wound rotor motor using simplified frequency relay system.

are used, the secondary of the motor cannot be shunted out completely. If the secondary can be completely removed from the circuit, the electron flow will take the path of least resistance, resulting in no energy being delivered to coils (A and B) upon starting.

The controllers for large crane hoists have a resistance, capacitance, and inductance control circuit network that is independent of the secondary rotor resistors.

Frequency relays have a number of advantages, including (1) positive response; (2) operating current drops sharply as the frequency drops below the point of resonance; (3) accuracy is maintained because this type of relay operates in a resonant circuit; (4) simple circuit; (5) changes in temperature and variations in line voltage do not affect the relay; and (6) an increase in motor load prolongs the starting time.

STUDY/DISCUSSION QUESTIONS

1. Are the secondary resistors connected in three uniform wye or delta sections? *THree (3) uniform wye sections.*

2. Do secondary resistors on starters with three or more steps of acceleration have more or less current inrush than those with two steps of acceleration? *Less   150%*

3. Does reversing the secondary rotor leads mean that the direction of rotation will reverse? *Lines 1&3 on Primary side interchanged.*

4. If one of the secondary resistor contacts (S2) fails in figure 32-1, what will happen? *motor would not acelerate to its full-speed.*

5. In figure 32-2, how many different interlocking conditions exist? *Parallel Limit Switches INTeRLock F&R Contact*

6. Referring to figure 32-4, why isn't it possible to remove all of the resistance from the secondary circuit? *Because normally closed contacts are used, the secondary of the motor cannot be shunted out completely.*

7. If frequency relays are used on starting, why is the starting cycle prolonged with an increase in motor load? *Because the motor will acelerate slower for a heavy load.*

8. If there is a locked rotor in the secondary circuit, what will be the value of the frequency? *% of the slip*

Unit 33 Automatic Speed Control

OBJECTIVES

After studying this unit, the student will be able to

- Identify terminal markings for wound rotor motors used with automatic speed controllers.

- Describe the purpose and function of automatic speed controllers for wound rotor motors.

- Connect and troubleshoot wound rotor motors and controllers for automatic speed control.

Automatic speed control of a wound rotor motor can be obtained with the use of pilot devices. The control of acceleration and deceleration, as well as the maintenance of selected speeds is possible. The line diagram of a controller using pilot devices to provide automatic speed control is shown in figure 33-1, page 126.

If it is assumed that the wound rotor motor in figure 33-1 is coupled to a fluid pump in a liquid-controlled system, then the operation of the system is as follows. To maintain the liquid level automatically, the selector switch is placed in the automatic position. As the liquid rises, the master float switch (MFS) closes the circuit to the control switch. As the fluid continues to rise, float switch (FS1) energizes the control relay (CR1). CR1 closes the main starter contacts (M) to start the motor in slow speed and energize the timing relay (T1). If the motor speed is too slow to permit proper fluid delivery, the changing liquid level in the tank eventually closes the third float switch (FS2). As a result, CR2 is energized through the now closed contacts of T1 to operate the first accelerating contactor (1A), the first bank of resistance is removed from the circuit, and the second time-delay relay (T2) is energized. This process continues until a motor speed is reached that maintains the liquid in the tank at a constant level. If the control selector switch is placed on manual, the motor must start with all of the resistance inserted in the secondary circuit. In addition, the motor must follow the preset timing sequence until all resistance is removed to obtain the maximum operation from the pump.

STUDY/DISCUSSION QUESTIONS

1. Referring to figure 33-1, can the motor be started in the manual position when there is no fluid in the tank? Why? *Yes, motor must start with all the resistance inserted in the secondary circuit.*

2. Assuming that this system is installed in an industrial mixing or processing tank, and that the tank input decreases, what actions will be initiated by the control system?

3. How does the connection of the overload relay contacts cause all other control relays and contactors to drop out? *Drops them out according to sequence.*

A third float switch, (FS2) closes CR2 is energized which operates. accelerating contactor (1A) second time-delay relay (T2) energizes motor speed is reached that maintains level at a constant

125

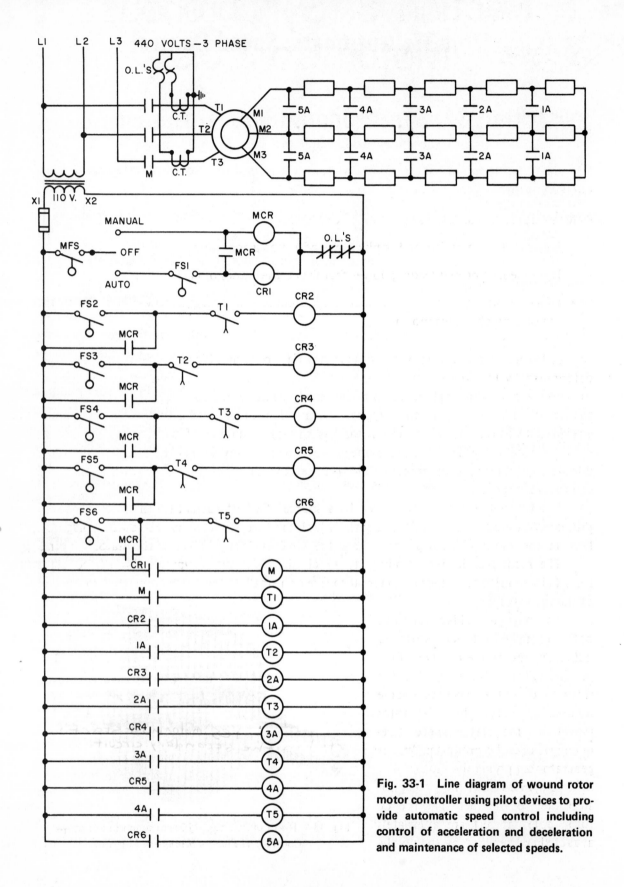

Fig. 33-1 Line diagram of wound rotor motor controller using pilot devices to provide automatic speed control including control of acceleration and deceleration and maintenance of selected speeds.

Section 8 Synchronous Motor Controls

Unit 34 Synchronous Motor Operation

OBJECTIVES

After studying this unit the student will be able to

- Describe the operation and applications of a synchronous motor.

- Describe lagging and leading power factor and the causes of each.

- Describe how to improve an electrical system having a lagging power factor with the use of a synchronous motor.

A distinguishing feature of the synchronous motor is that it runs without slip at a speed determined by the frequency and the number of poles it contains. This type of motor sets up a rotating field through stator coils energized by alternating current (this action is similar to the principle of an induction motor). An independent field is established by a rotor energized by direct current. The rotor has the same number of coils as the stator. At running speed, these fields lock into one another magnetically so that the speed of the rotors is in step with the rotating magnetic field of the stator. In other words, the rotor turns at the synchronous speed.

The rotor, figure 34-1, is excited by a source of direct current for the sole purpose of producing alternate north and south poles which then are attracted by the rotating magnetic field in the stator. The rotor must have the same number of poles as the stator winding.

The rotor has dc field windings to which direct current is supplied through collector rings (slip rings) from either an external source or a small dc generator connected to the end of the rotor shaft.

The magnetic fields of the rotor poles are locked into step with and pulled around by the revolving field of the stator. Assuming that the rotor and stator have the same number of poles, then the rotor moves at the stator speed (frequency) actually produced by the generator supplying the motor.

A synchronous motor cannot start due to the fact that the dc rotor poles at rest are alternately attracted and repelled by the revolv-

Fig. 34-1 Rotor of 300-h.p., 720-rpm synchronous motor. (Courtesy of Electric Machinery Mfg. Co.)

Fig. 34-2 2000-h.p., 300-rpm synchronous motor driving compressor

ing stator field. An induction or starting winding, therefore, is embedded in the pole faces of the rotor.

The starting winding resembles a squirrel cage winding. The induction effect of the starting winding provides the starting, accelerating, and pull-up torques required. This winding cannot be used like that of the conventional squirrel cage motor; it is designed to be used for starting and for damping oscillations during running only. It has a relatively small cross-sectional area, and thus will overheat if the motor is used as a squirrel cage induction motor.

The slip is equal to 100 percent at the moment of starting. Thus, the action of the ac rotating magnetic field of the stator cutting the rotor windings (which are stationary at startup) may produce voltages high enough to damage the insulation if precautions are not taken.

If the dc rotor field is either connected as a closed circuit or connected to a discharge resistor during the starting period, the resulting current produces a voltage drop that is opposed to the generated voltage. As a result, the induced voltage at the field terminals is reduced. The squirrel cage winding is used to start the synchronous motor in exactly the same way as in the squirrel cage induction motor. When the rotor reaches the maximum speed to which it can be accelerated as a squirrel cage motor (about 95 percent or more of the synchronous speed), direct current is applied to the rotor field coils to establish north and south rotor poles. These poles then are attracted by the poles on the stator and cause the rotor to accelerate until it locks into synchronous motion with the stator field.

Synchronous motors are used for applications involving large, slow-speed machines with steady loads and constant speeds, such as compressors, fans and pumps, many types of crushers and grinders, and pulp, paper, rubber, chemical, flour, and metal rolling mills, figure 34-2.

POWER FACTOR CORRECTION BY SYNCHRONOUS MOTOR

The question of power factor is of great concern to the industrial user of electricity. Power factor is the ratio of the actual power (expressed in watts or kilowatts, kw) being used in a circuit to the power apparently being drawn from the line, expressed in volt-amperes or kilovolt-amperes (kva). The kva value is obtained by multiplying a voltmeter reading and an ammeter reading of the same circuit or equipment. Inductance within the circuit will cause the current to lag the voltage.

When the values of the apparent and actual power are equal or in phase, the ratio of these values is 1:1. In other words, when the voltage and amperage are in phase, the ratio of these values is 1:1, such as in the case of pure resistive loads. This power factor value of unity is the highest power factor that can be obtained. The higher the power factor, the greater is the efficiency of the electrical equipment.

Ac loads generally have a lagging power factor. As a result, these loads burden the power system with a large reactive load. A synchronous motor with an overexcited dc field may be used to offset the low power factor of the other loads on the same electrical system. An overexcited synchronous motor means that it is operating at more than the unity power factor.

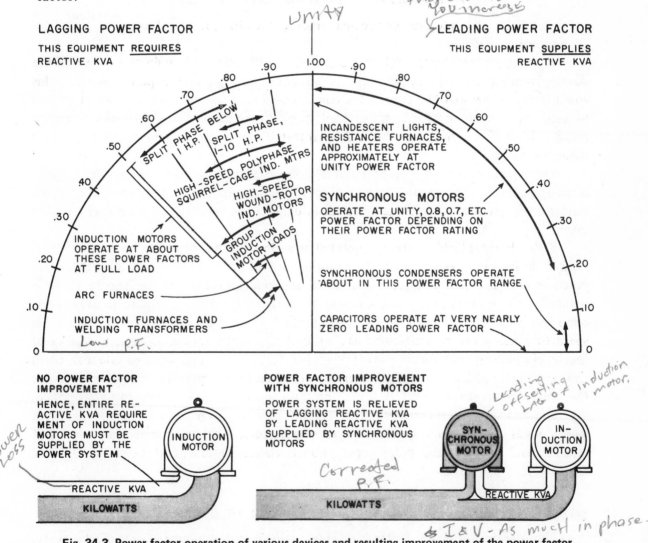

Fig. 34-3 Power factor operation of various devices and resulting improvement of the power factor operation with the use of synchronous motors. (Courtesy Electric Machinery Mfg. Co.)

STUDY/DISCUSSION QUESTIONS

1. Why is it necessary that the rotor and stator have an equal number of poles? *The rotor must have same number of poles as the stator windings so that synchronous speed will be obtained.*

2. What is the effect of the starting winding of the synchronous motor on the running speed? *Induction effect*
   Dampers

3. What are typical applications of synchronous motors?
   Compressors, Fans, & Pumps

D.C. Exciter
4. Explain how the rotor magnetic field is established. *Locked into step*
   with revolving mag. field of stator

5. A loaded synchronous motor cannot operate continuously without dc excitation on the rotor. Why? *Producing alternate north and south poles*
which attract a rotating magnetic field in the stator

6. Why must a discharge resistor be connected in the field circuit for starting? *Induced voltage at field terminals are reduced. Across Rotor field*

7. What is meant when a synchronous motor is called overexcited? *Operating at more than unity power factor*

8. Depending on their power factor rating, what is the range of leading power factor at which synchronous motors operate? *Operate at Unity*
   0.8   0.7

9. At what power factor do incandescent lights operate? *Unity Power factor*
   factor

10. At what power factor do high-speed, wound rotor motors operate?
    .80 to .90 Lagging

Dampers
START only

Unit 35 Pushbutton Synchronizing

OBJECTIVES

After studying this unit, the student will be able to

- Identify the terminal markings for a pushbutton synchronizing controller and a synchronous motor.
- Describe the two functions of synchronous motor control.
- Connect and troubleshoot synchronous motors and synchronizing controls.
- Describe the adjustments that can be made to the circuit of a synchronous motor and control to obtain a power factor of unity or greater.

There are two basic functions of synchronous motor control.

1. Synchronous motor control means that the motor can be started as an induction motor. As a result, the motor can be started using any of the typical induction motor starting schemes, including connecting the motor across the line or the use of autotransformers, primary resistors, or other devices.

2. Synchronous motor control also brings the motor up to synchronous speed by exciting the dc field. The principal difference between synchronous motor control and induction motor control is in the control of the field.

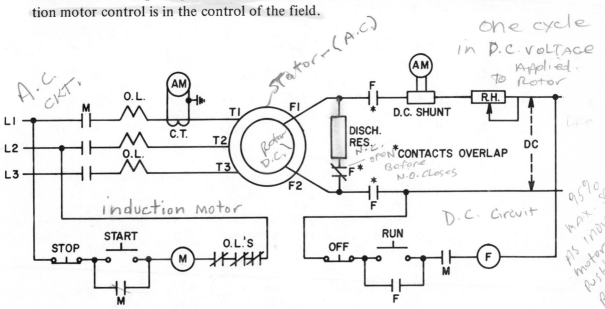

Fig. 35-1 Elementary diagram of pushbutton synchronizing.

In figure 35-1, the synchronous motor is started as an induction motor by pressing the start button. When the motor reaches its maximum speed, the run pushbutton is closed to energize coil (F) and close the dc excitation contacts (F). These contacts, in turn, open the field discharge N. C. contact (F), and energize the field.

131

The ammeter and rheostat shown in the dc circuit in figure 35-1 provide control of the excitation current. The unity power factor of the motor can be found by adjusting the rheostat to obtain a minimum reading on the ac ammeter. If the dc field is excited to a greater degree, a leading power factor is created which is helpful to a lagging distribution system.

Adjustments of the circuit may cause field currents, or line currents, in excess of the motor rating. As a result, electrical instruments must be provided to monitor various circuit values so that the operator can prevent equipment stoppage due to overload tripping. The readings of the instruments must not exceed the rated values shown on the motor nameplate.

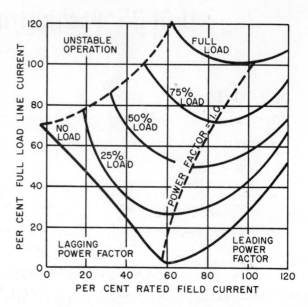

Fig. 35-2 Line current vs. field current for a motor with a unity power factor.

The excitation current can be applied manually to excite the dc field to bring the motor up to synchronous speed. For this operation, however, it is the operator who must make the difficult judgment of when the north and south poles of the rotor and stator fields are paired and the motor is ready for synchronous operation.

STUDY/DISCUSSION QUESTIONS

1. What are the two basic functions of synchronous motor control?
   1.) induction motor start
   2.) synchronous speed
2. What is the purpose of the interlock (M) in the dc control circuit?
   Closes and Energizes F coil when motor up to full speed
3. If all of the control circuits are ac and dc is not available, what will be the result of energizing all of the contactor coils? *there would be no synchronous control available.*
4. When the rotor and stator fields lock into step, why may the reading of the ac ammeter decrease? *voltage drop opposed to generated voltage*
5. If the dc field becomes overexcited (power factor greater than unity) the ac ammeter may increase its scale reading. Why? *I+ will help a lagging distribution system.*
6. Why is manual field application not very satisfactory? *The operator must make the critical judgement of when N AND S poles of rotor and stator fields are paired*
7. If a synchronizing attempt fails, what procedure should be followed? *Press stop button and repeat the starting cycle*
8. In figure 35-2, approximately what is the percentage of current drawn from the line when the field current is adjusted to 100 percent of its capacity at full motor load? *100%*
9. Why are the ac ammeter and dc meter necessary in the circuit of figure 35-1? *provide control of of the excitation current.*
10. Why is the operation of the field contacts (F) overlapped in figure 35-1?
    Because of Sequence

Unit 36 Timed Semiautomatic Synch

OBJECTIVES

After studying this unit, the student will be able to

- Identify terminal markings for a timed semiautomatic controller used with a sy̶ nous motor.

- Describe the operation of a timed semiautomatic controller in bringing a motor up to synchronous speed.

- Connect and troubleshoot synchronous motors and timed semiautomatic controllers.

A synchronous motor may be brought up to synchronous speed with the use of a definite time relay to excite the field time relay. This method is shown in the circuit of figure 36-1.

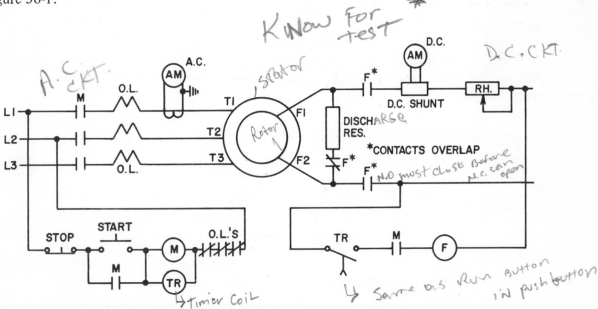

Fig. 36-1 Elementary diagram of a timed, semiautomatic synchronizing installation.

The timing relay (TR) is energized with the main starter coil (M). The instantaneous N. O. interlock (M) is then closed. Both interlock M and the dc contactor coil (F) must wait for the closing of the delay-in-closing contact (TR). After a preset timing period, the rotor has accelerated to the maximum speed possible at this stage. Contact (TR) then closes to accelerate the rotor until it reaches the point where it synchronizes. The timer setting should be adjusted for the maximum time required to accelerate the motor to the point that it can reach the synchronous speed after contact (TR) closes.

The attempt to synchronize the motor may not be successful. Then it is necessary to press the stop button and repeat the starting cycle. It is not necessary to bring the rotor to a standstill; only the timing cycle must be reactivated.

he equipment operator and the electrician should realize that both methods of syn-
ous motor control (pushbutton control, unit 35, and timed semiautomatic control) are
t guaranteed to be effective on every attempt to obtain synchronous operation of the
motor.

STUDY/DISCUSSION QUESTIONS

1. What assurance is there that the rotor will lock into step at the synchronous speed with the use of a timing relay? Energized with main
   Starter coil.   None

2. If the motor fails to achieve synchronous operation, what action is
   necessary? press stop button and repeat
   Starting cycle.

Unit 37 Synchronous Motor Starter with Polarized Field Frequency Relay

OBJECTIVES

After studying this unit, the student will be able to

- Describe how an out-of-step relay protects the starting winding of a synchronous motor.
- Describe the action of a polarized field frequency relay in applying and removing dc field excitation on a synchronous motor.
- Connect and troubleshoot synchronous motors and controllers which use out-of-step relays and polarized field frequency relays to achieve automatic motor synchronization.

A polarized field frequency relay can be used for the automatic application of field excitation to a synchronous motor. There are two basic methods of starting synchronous motors. In the first method, full voltage is applied to the stator winding, and in the second method, the starting voltage is reduced. The most commonly used of these two methods of starting synchronous motors is the across-the-line connection in which the stator of the synchronous motor is connected directly to the plant distribution system at full voltage. A magnetic starter is used in this method of starting.

ROTOR CONTROL EQUIPMENT

Field Contactor

The field contactor opens both lines to the source of excitation. During starting, the contactor also provides a closed field circuit through a discharge resistor. A solenoid-operated field contactor is very similar in appearance to the standard dc contactor. However, for the solenoid-operated contactor, the center pole is normally closed and designed to provide a positive overlap between the normally closed contact and the two normally open contacts. This overlap is an important feature, since at no time should the field winding be open. The field winding of the motor must always be short circuited through a discharge resistor or connected to the dc line. The coil of the field contactor is operated from the same direct-current source that provides excitation for the synchronous motor field.

Out-of-Step Relay

If a synchronous motor starts, accelerates, and reaches synchronous speed within a time

Fig. 37-1 Typical magnetic contactor used on synchronous starters for field control (Courtesy Allen-Bradley Co.)

interval determined to be normal for the motor, and if the motor continues to operate at synchronous speed, then the squirrel cage or starting winding will not overheat. Under these conditions, adequate protection for the entire motor is provided by three overload relays in the stator winding. The squirrel cage winding, however, is provided for starting purposes only. If the motor operates at subsynchronous speed, the squirrel cage winding may sustain damage due to overheating. It is not unusual for some synchronous motors to withstand a maximum locked rotor interval of only five to seven seconds.

To protect the starting winding, an out-of-step relay (OSR), figure 37-2, is provided on synchronous starters. If the motor does not accelerate and reach the synchronizing point after a preset time delay (or return to a synchronized state after leaving it), and if the amount of current induced in the field winding exceeds a value determined by the core setting of the out-of-step relay, then the normally closed contacts of the relay will open to deenergize the line contactor. As a result, power is removed from the stator circuit before the motor overheats.

Fig. 37-2 Out-of-step relay used on synchronous starters. (Courtesy Allen-Bradley Co.)

 Polarized Field Frequency Relay

The process of starting a synchronous motor is simply the acceleration of the motor to as high a speed as possible from the squirrel cage winding and then the application of the dc field excitation. The component which is responsible for correctly and dependably applying and removing the field excitation is the polarized field frequency relay and reactor, figure 37-3.

Fig. 37-3 Polarized field frequency relay with contacts in normally closed position. (Courtesy Allen-Bradley Co.)

The operation of the frequency relay is shown in figure 37-4. The magnetic core of the relay has a direct-current coil (C), an induced field-current coil (B), and a pivoted armature (A) to which contact (S) is attached. Coil (C), connected to the source of dc excitation, establishes a constant magnetic flux in the relay core. This flux causes the relay to be polarized. Superimposed on this magnetic flux in the relay core is the alternating magnetic flux produced by the alternating induced current flowing in the coil (B). The flux through armature (A) depends on the flux produced by coils (B) and (C). Coil (B) produces an alternating flux of equal positive and negative magnitude each one-half cycle. Thus, the combined flux flowing through armature (A) is much larger when the flux from coil (B) opposes that from coil (C). In figure 37-4A, the flux from coil (B) opposes the flux from dc coil (C), with the result that a strong resultant flux is forced through armature (A) of the relay. This condition

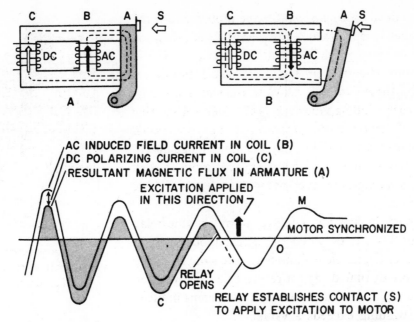

Fig. 37-4 Polarized field frequency relay operation (Courtesy Electric Machinery Mfg. Co.)

is shown by the lower shaded loops of figure 37-4C. One-half cycle later, the flux produced by coil (B) reverses and less flux flows through armature (A). This is due to the fact that the flux from coil (B) no longer forces as much flux from coil (C) to take the longer path through armature (A). The resultant flux is weak and is illustrated by the small, upper shaded loops of figure 37-4C. The relay armature opens only in the span of the induced field current wave represented by the small, upper loops of the relay armature flux.

As the motor reaches synchronous speed the induced field current in relay coil (B) decreases in amplitude, and a value of relay armature flux (upper shaded loop) is reached at which the relay armature (A) no longer stays closed. The relay then opens to establish contact (S) which allows dc excitation to be applied at the point indicated on the induced field current wave, figure 37-4C.

Excitation is applied in the direction shown by the arrow. The excitation is opposite in polarity to that of the induced field current at the point of application. This procedure is necessary to compensate for the time required to build up excitation. The time interval results from the magnetic inertia of the motor field winding. Because of the inertia, the dc excitation does not become effective until the induced current reverses (point O on the wave) to the same polarity as the direct current. The excitation continues to build up until the motor is synchronized as shown by point M on the curve.

Figure 37-5 indicates the normal operation of the frequency relay. Dc excitation is applied to the coil of the relay at the instant the synchronous motor is started. When the stator winding is energized using either full voltage or reduced voltage methods, line current is allowed to flow through the three overload relays and the stator winding. As a result, line frequency currents are induced in the two electrically independent circuits of the rotor: the squirrel cage or starting windings, and the field windings. The current induced in the field windings flows through the reactor which then shunts part of it through the ac coil of the frequency relay, the coil of the out-of-step relay, the field discharge resistor, and finally to

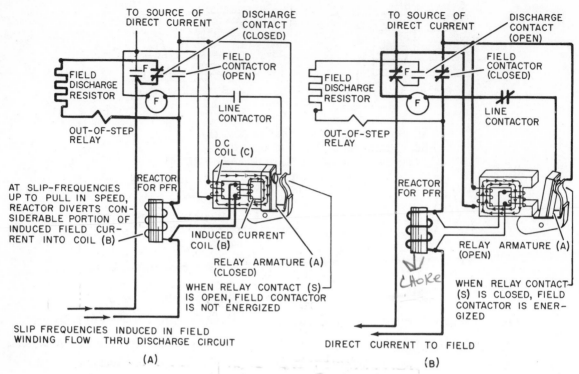

Fig. 37-5 Wiring connections and operation of a polarized field frequency relay.
(Courtesy Electric Machinery Mfg. Co.)

the normally closed contact of the field contactor. The flux established in the frequency relay core pulls the armature against the spacer and opens the normally closed relay contacts, figure 37-5. As the motor accelerates to the synchronous speed, the frequency of the induced currents in the field windings diminishes. There is, however, sufficient magnetic flux in the relay core to hold the armature against the core. This flux is due to the fact that the impedance of the reactor at high slip frequency forces a considerable amount of induced current through the ac coil of the frequency relay.

At the point where the motor reaches its synchronizing speed, usually 92 to 97 percent of the synchronous speed, the frequency of the induced field current is at a very low value. The reactor impedance also is greatly reduced at this low frequency. As a result, the amount of current shunted to the ac coil is reduced to the point where the resultant core flux is no longer strong enough to hold the armature against the spacer. At the precise moment that the rotor speed and the frequency and polarity of the induced currents are most favorable for synchronization, the armature is released, the relay contacts close, and the control circuit is completed to the operating coil of the field contactor. Thus, dc excitation is applied to the motor field winding, figure 37-5B. At the same time, the out-of-step relay and discharge resistor are deenergized by the normally closed contacts of the field contactor.

An overload or voltage fluctuation may cause the motor to pull out of synchronism. In this event, current of slip frequency is induced in the field windings. Part of this current flows through the ac coil of the polarized field frequency relay, opens the relay contact, and removes the dc field excitation. If line voltage and load conditions return to normal within a preset time interval, and the motor has enough pull-in torque, then the motor

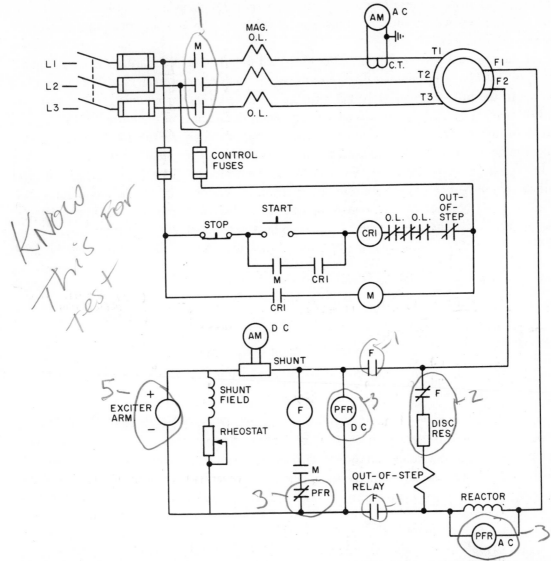

Fig. 37-6 Line diagram for automatic operation of synchronous motor using polarized field frequency relay.

automatically resynchronizes. However, if the overload and low-voltage conditions continue so that the motor cannot resynchronize, then either the out-of-step relay or the overload relays activate to protect the motor from overheating.

SUMMARY OF AUTOMATIC STARTER OPERATION

In figure 37-6, a line diagram is shown for the automatic operation of a synchronous motor. For starting, the motor field winding is shorted through the normally closed power contact of the field contactor (F), the discharge resistor, the coil of the out-of-step relay, and the reactor. When the start button is pressed, the circuit is completed to the pilot contactor (CR1) through the control fuses, the stop button, and contacts of the overload and out-of-step relays. The closing of CR1 energizes the line contactor (M) which applies full voltage at the motor terminals with the overload relays in the circuit. A normally open contact on

CR1 and a normally open interlock on the line contactor (M) provide the hold-in circuit. An ammeter, supplied with a current transformer, indicates the starting and running current drawn by the motor.

At the moment the motor starts, the polarized field frequency relay opens its normally closed contact and maintains an open circuit to the field contactor (F) until the motor accelerates to the proper speed for synchronizing. When the motor reaches a speed equal to 92 to 97 percent of its synchronous speed, and when the rotor is in the correct position, the contact of the polarized field frequency relay closes to energize the field contactor (F) through an interlock on the line contactor (M). The closing of the field contactor (F) applies the dc excitation to the field winding and causes the motor to synchronize. After the field circuit is established through the normally open power contacts of the field contactor, the normally closed contact on this contactor opens the discharge circuit. The motor is now operating at the synchronous speed. If the stop button is pressed or if either magnetic overload relay is tripped, the starter is deenergized and disconnects the motor from the line.

STUDY/DISCUSSION QUESTIONS

1. What are the two basic methods of starting a synchronous motor?
   Across Line (Full Voltage) or Reduced Voltage

2. What is an out-of-step relay?
   SCR to Long to come up to speed

3. Why is an out-of-step relay used on synchronous starters?
   Protect Starting Winding

4. Under what conditions will the out-of-step relay trip out the control circuit? *TAkes to Long to synchronize*

5. What is the last control contact which closes on a starting and synchronizing operation? *FC contact*

6. What influence do both of the polarized field frequency relay coils exert on the N. C. contact (PFR)? *magnetic flux to open contacts*

7. Why is the PFR relay polarized with a dc coil? *proper synch timing between A.C. & D.C.*

8. Approximately how much time (in terms of electrical cycles) elapses from the moment the PFR relay opens to the moment the motor actually synchronizes? *one cycle*

9. How does the ac coil of the PFR relay receive the induced field current without receiving the full field current strength? *thru Reactor*

10. Why is a control relay (CR1) used in figure 37-6?
    To energize Line contactor (M)
    more suseptive to U Voltage protection

Poles of Rotor same As STATOR

Section 9 Direct-Current Controllers

Unit 38 Control Relays

OBJECTIVES

After studying this unit, the student will be able to

- Describe the operation of a mechanical latch relay.
- Describe the operation of a thermostat relay used with a three-wire control device such as a thermostat.
- Connect and troubleshoot circuits using thermostat relays.

MECHANICAL LATCH RELAY

A direct-current mechanically held relay is used in a manner similar to that for alternating-current mechanically held relays (Unit 5). A current is not required to hold the contacts in the closed position. The relay is closed by momentarily energizing the coil; it is then held closed by a mechanical latch. When a second coil is momentarily energized the latch is tripped and the relay opens.

THERMOSTAT RELAY

Thermostat-type relays, figure 38-3, are used with three-wire, gauge-type thermostat controls or other pilot controls having a slowly moving element which makes a contact for both the closed and open positions of the relay. The contacts of a thermostat control device usually cannot handle the current to a starter coil; therefore, a thermostat relay must be used between the thermostat control and the starter. This application of a thermostat relay and its related circuit is also valid for use with ac.

When the moving element of the thermostat control touches the closed contact, the relay closes and is held in this position by a maintaining contact. When the moving element touches the open contact, the current flow bypasses the operating coil through a small resistor and thus causes the relay to open. The resistor is usually built into the relay and serves to prevent a short circuit.

The thermostat contacts must not overlap or be adjusted too close to one another as this may result in the resistance

(Courtesy Allen-Bradley Co.)

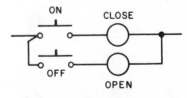

Fig. 38-1 Mechanical latch relay (top) and electrical control connections for operation.

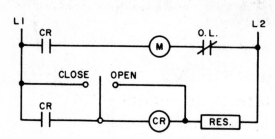

Fig. 38-2 Starter coil (M) is controlled by thermostat relay.

Fig. 38-3 Thermostat relay (Courtesy Allen-Bradley co.)

unit being burned out. It is also advisable to compare the inrush current of the relay with the current rating of the thermostat.

STUDY/DISCUSSION QUESTIONS

Relay closed by
mech. latch →

1. What are the advantages of a mechanically held relay? *Current not*
   Required to hold contacts in closed position.

2. Why does coil (CR) in figure 38-2 drop out when the thermostat touches the open position?

By-passes operating
coil.

Unit 39   Across-the-Line Starting

OBJECTIVES

After studying this unit, the student will be able to

- Describe one method of across-the-line starting for small dc motors.

- State why a current limiting resistor may be used in the starting circuit for a dc motor.

- Connect and troubleshoot across-the-line starters used with small dc motors.

Small dc motors may be started by connecting them directly across the line. Fractional horsepower manual starters (unit 2) or magnetic contactors and starters (unit 3) are used for across-the-line motor starting, figure 39-1.

Magnetic across-the-line control of small dc motors is similar to ac control or to two- or three-wire control. However, due to the added load of multiple contacts and the fact that the dc circuit lacks the inductive reactance which is present with the ac electromagnets, some dc across-the-line starter coils have dual windings. Both windings are used to lift and close the contacts, but only one winding remains in the holding position. The starting winding, or lifting winding, of the coil is designed for momentary duty only. In figure 39-2, assume that coil (M) is energized momentarily. When the starter is closed, it maintains itself through the N. O. maintaining contact (M) and the upper winding of the coil since the N. C. contact (M) is now open. Power contacts (M) close, and the motor starts across the full line voltage.

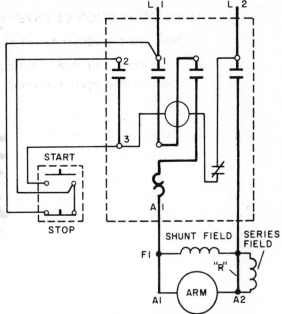

Fig. 39-1 Dc full voltage starter wiring diagram. Connection ("R") is removed with use of series field.

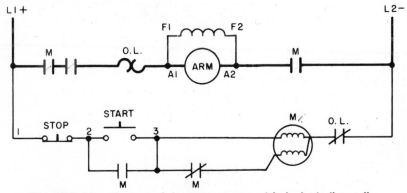

Fig. 39-2 Line diagram of dc motor starter with dual winding coil.

143

Figure 39-3 shows another method used to start a dc.motor. In this method, a current limiting resistor is used when it is necessary to limit a continuous duty current flow to some coils or when it is found that coils are overheating.

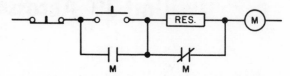

Fig. 39-3 Dc starting circuit using current limiting resistor.

The coil receives the maximum current required to close the starter and then the minimum current necessary to hold in the contacts and for continuous duty through the current limiting resistor.

STUDY/DISCUSSION QUESTIONS

1. Why may small dc motors be started directly across the line? DUAL WINDINGS COIL.
2. What may happen if a resistor is not added to the circuit for continuous duty when the coil is not designed for continuous duty?

Unit 40 Use of Series Starting Resistance

OBJECTIVES

After studying this unit, the student will be able to

- List the two factors which limit the current taken by a motor armature.
- State why it is necessary to insert resistance in part of a dc motor circuit on starting.
- Calculate the amount of series resistance required for a dc motor at different speeds during the starting period.

The starting of fairly large dc motors requires that resistance be inserted in series with the armature of the motor.

Two factors which limit the current taken by a motor armature from a given line are the counter emf and the armature resistance. Since there is no counter emf when the motor is at a standstill, the current taken by the armature will be abnormally high unless an external current limiting resistor is used.

Figure 40-1 illustrates a shunt motor connected directly across a 250-volt line. The resistance of the armature is 0.5 ohm. The full-load current taken by the motor is 25 amperes and the shunt field current is one ampere. The resulting armature current at full load is 24 amperes.

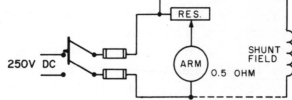

Fig. 40-1 This 25-ampere dc motor will draw 500 amperes from the line on starting.

It was stated in a previous unit that the counter emf developed by a motor is proportional to the speed of the motor if the field is constant. In addition, the counter emf equals the applied voltage minus the armature (IR) drop. The current through the armature is determined by the equation:

$$I = \frac{E - E_c}{R_a}$$

where: I = armature current in amperes
E = line voltage in volts $250\ V$
E_c = counter emf in volts $E_c = -(I \times R_A)$
R_a = armature resistance in ohms $.05$

Therefore, the current through the armature at the instant of starting for the shunt motor in figure 40-1 is:

$$I = \frac{250 - 0}{0.5} = 500 \text{ amperes} \quad +1\ \text{AMP SHUNT FIELD}$$

The total starting current is equal to the armature current plus the field current of one ampere. Thus, the ratio of the starting current to the full load current is:

$$\frac{501}{25} = 20.04$$

$\longrightarrow$ FULL LOAD CURRENT TAKEN BY MOTOR

The excessive torque and heat produced by this current may harm the motor and the load attached to it, as well as causing the insulation to overheat on starting and the armature to burn up.

These undesirable effects can be eliminated by connecting a resistance in series with the armature to reduce the starting armature current to approximately 1.5 times the full-load current value and then gradually removing the resistance from the circuit.

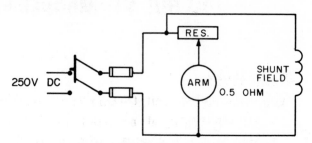

Fig. 40-2 **Starting resistors are used to eliminate high starting currents.**

The value of this series resistance may be obtained by solving the previous equation for (R) (remember that the armature resistance (R_a) must be subtracted):

$$R = \frac{E - E_c}{I} - R_a$$

When the motor is at a standstill the series resistance is equal to:

$$R = \frac{250 - 0}{36} - 0.5 = 6.44 \quad \text{ohms}$$

The required series resistance also may be determined for some intermediate speed in the starting period of the motor. In this case, it is necessary to solve for the counter emf at the speed in question and then obtain the resistance from the above formula. Since the counter emf differs from the applied voltage only by the armature (IR) drop, then the speed is nearly proportional to the applied voltage. For example, at half speed, the counter emf is equal to:

$$\frac{50 \% \text{ speed}}{100 \% \text{ speed}} = \frac{\text{Voltage at Half Speed}}{250}$$

$$\text{Voltage at Half Speed} = 250 \times \frac{50\%}{100\%} = 250 \times 1/2 = 125 \text{ volts}$$

The resistance required at half speed is equal to:

$$\frac{250 - 125}{36} - 0.5 \text{ OR } 2.97 \text{ ohms}$$

The counter emf can be found at the full load:

$$E_c = E - (I\,R_a)$$
$$E_c = 250 - (24 \times 0.5)$$
$$E_c = 250 - 12$$
$$E_c = 238 \text{ volts}$$

STUDY/DISCUSSION QUESTIONS

1. Why is a starter necessary on a dc motor?
2. Why is a resistance placed in series with the armature only, rather than in series with the entire shunt motor?
3. If an armature has a resistance of 0.5 ohms, what series resistance is necessary to limit the current to 30 amperes across a 230-volt circuit at the instant of starting? What is the power rating?
4. What resistance is necessary when the motor in question 3 reaches 50% of its rated speed?

Unit 41 Manual Faceplate Starters

OBJECTIVES

After studying this unit, the student will be able to

- Identify terminal markings for dc shunt motors and manual faceplate starters.
- Describe the operation of a three-terminal dc manual faceplate starter.
- Connect and troubleshoot direct current motors and manual faceplate controllers.
- Describe the difference between a three-terminal and a four-terminal dc manual starter.

THREE-TERMINAL DC FACEPLATE STARTER

A three-terminal, dc manual starter is a device that contains the series resistors required for starting a motor. It also contains the contacts and switching arm required to transfer through these resistors. In addition, the manual starter contains a device which reinserts this resistance in the circuit automatically in the event of a line voltage failure.

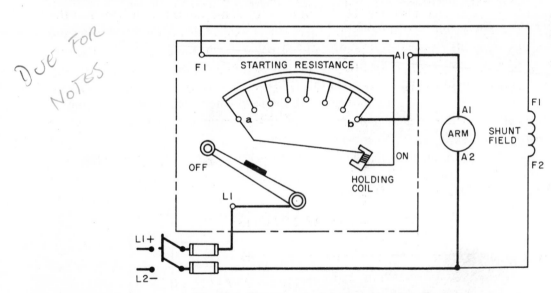

Fig. 41-1 Three-terminal dc manual starter diagram.

Figure 41-1 represents a three-terminal starter connected to a shunt motor. The direct-current line is connected to the starter and motor through a switch and suitable fuses. When the contact arm is moved from the off position to the first contact (a) of the starting resistance, the armature, which is in series with the total starting resistance, is connected across the line; the shunt field, in series with the holding coil, is also connected across the line. As a result of this method of operation, the initial current inrush to the armature is limited by the resistance to a reasonable value, and the field current is at the maximum value to provide a good starting torque.

As the contact arm is moved to the right, the starting resistance is reduced and the motor accelerates. When the last contact (b) is reached, the armature is connected directly across the line, and the motor is at full speed.

147

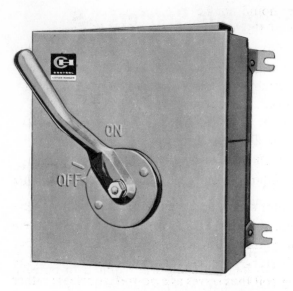

Fig. 41-2 Three-terminal, manual faceplate starter with armature resistor;
shown with cover (left) and with cover removed. (Courtesy Cutler-Hammer, Inc.)

The holding coil is connected in series with the shunt field to provide a no field release. In the event the field circuit opens, the motor speed will become excessive if the armature remains connected across the line. To prevent this if there is a field failure, the holding coil in figure 41-1 is demagnetized and the arm returns to the off position.

Note that the starting resistance is in series with the field when the arm is in the running position. The effect of this resistance on the speed of the motor is negligible because the starting resistance is only a small percentage of the shunt field resistance. The current and voltage drop, therefore, are small.

When this type of starter is operated, the arm must be moved slowly to limit the current inrush to the motor. The arm should not be held too long on any contact between (a) and (b) since the starting resistance is designed to carry the starting current for a short time only.

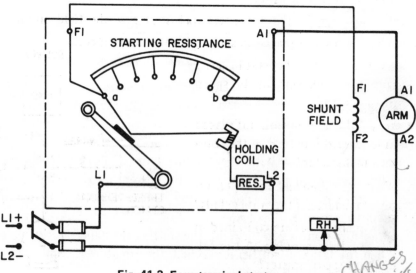

Fig. 41-3 Four-terminal starter

In other words, the speed of the motor should not be controlled by holding the arm for any length of time on an intermediate position between contacts (a) and (b).

If a motor has an external resistance in series with the field to provide speed control, then the three-terminal starter is not a suitable choice for operating the motor. Since the external resistance may reduce the current in the field and holding coil to a value insufficient to hold the arm against the action of its spring, the arm will return to the off position.

Fig. 41-4 Circular plate rheostat. (Courtesy Cutler-Hammer, Inc.)

FOUR-TERMINAL DC FACEPLATE STARTERS

In applications where a large variety of motor speeds are required, a four-terminal faceplate starter is used on the motor. The four-terminal starter, figure 41-3, differs from the three-terminal starter in that the holding coil is not connected in series with the shunt field but is connected across the line in series with a resistor. This resistor limits the current in the holding coil to the desired value. The holding coil thus serves as a no-voltage release rather than as a no-field release. If the line voltage drops below the desired value, the attraction of the holding coil is decreased, and the spring pulls the arm back to the off position.

The four-terminal starter is used when it is desired to increase the speed of the motor by inserting a field rheostat to weaken the field current. The motor is started by slowly moving the arm from (a) to (b), where it is held by the holding coil. The speed of the motor is controlled by varying the field rheostat. The speed is at a minimum when the resistance is cut out entirely from the circuit and at a maximum when all of the resistance is included in the circuit.

STUDY/DISCUSSION QUESTIONS

1. Why is the holding coil used?
2. Why is the resistance cut out in steps rather than all at once?
3. What will happen if the starting arm is held on a contact between (a) and (b) for a long time?
4. What occurs in the three-terminal starter if the field circuit opens?
5. Why is a resistor placed in series with the holding coil?
6. Explain how the rheostat in series with the field controls the speed of the motor.
7. Why is the holding coil not placed in series with the field in the four-terminal starter?
8. What will happen in the four-terminal starter if the voltage fails?
9. How many wires are contained in the conduits running to the following type of starter?

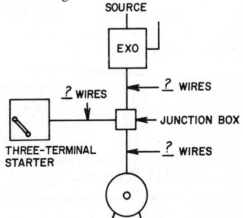

Unit 42 Counter EMF Controller

OBJECTIVES

After studying this unit, the student will be able to

- Identify terminal markings for a direct-current compound motor and a counter emf controller.
- Describe the operation of a counter emf controller.
- State the purpose of a relay connected across the armature of a counter emf controller.
- Connect and troubleshoot dc motors and counter emf controllers.

At the moment a dc motor starts, the counter emf across the armature is low. As the motor accelerates, the counter emf increases. When the voltage across the motor armature reaches a certain value, a relay is actuated to reduce the starting resistance at the right time.

Figure 42-1 shows the schematic connection diagram for a dc motor controller which uses the principle of counter emf acceleration. When the start button is pressed, contactor (M) is energized and the main contact (M) is closed. The current path is then complete from L1 through the resistor (R1-R2), the armature, and the series field to L2.

The motor will stop when the start button is released due to the fact that when M is

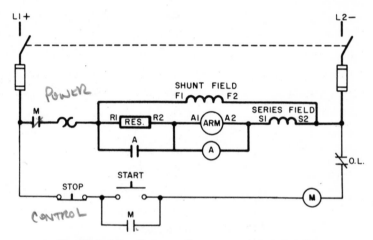

Fig. 42-1 Line diagram of a counter emf controller.

deenergized, the main contactor (M) opens. To prevent the contactor from opening, the maintaining contact (M) is operated from the coil (M), which places a shunt around the start button. As a result, when the start button is released M remains energized through the auxiliary relay contacts (M) until the stop button is pressed.

There is a high voltage drop across R1-R2, but a counter emf is not present across the armature. As the motor accelerates, the current is reduced in value. When the counter emf across the armature reaches a value equal to a certain percent of the full value, the relay (A) (connected across the armature) closes contact (A) with the result that the resistor (R1-R2) is removed from the circuit and the motor is placed directly on the line.

This type of counter emf controller automatically adjusts the starting time intervals, depending upon the load connected to the motor. For example, since a heavy load has more inertia for the motor to overcome, a longer time is required to build up the counter emf, and a longer starting period is required.

STUDY/DISCUSSION QUESTIONS

1. Why is this type of controller called a counter emf controller?

2. What physical characteristic distinguishes between a *main contactor* and a *relay*?

3. What is one method of increasing the motor speed above the base speed?

CONTROLLER - HANDLE
CURRENT FOR
VARIABLE AMOUT OF
TIME

Unit 43 Magnetic Time Limit Controller

OBJECTIVES

After studying this unit, the student will be able to

- Identify terminal markings for direct-current, compound motors and magnetic time limit controllers.

- Describe the sequence of operation of a magnetic time limit controller.

- Connect and troubleshoot dc compound motors and controllers for magnetic time limit acceleration.

The magnetic time limit controller provides time delay by causing the magnetic flux of a coil to decay slowly. A copper sleeve surrounding the iron core is used to obtain this slow decay in coil flux.

When the current in the main coil decreases, the decreasing flux induces a current in the short circuited sleeve. This current, in turn, produces a flux which tends to prolong the time period over which the coil retains control of its contactor. The time setting of the contactor depends upon the pressure of the spring that pulls the contactor open.

The elementary diagram for a magnetic time limit starter, figure 43-1, shows that when the start button is pressed, a current path is established from L1, through the stop button (normally closed), the start button, the normally closed interlock (M), and coil (A) to L2. Coil (A) is energized and thus opens the normally closed contacts (A) and closes the normally open interlock (A).

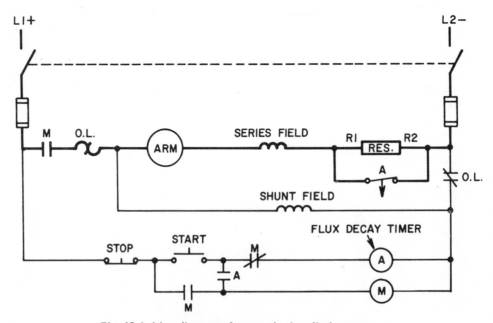

Fig. 43-1 Line diagram of magnetic time limit starter.

Current can then pass from L1 through the stop button, the start button, A (which has been closed), and M to L2. Since coil (M) is now energized, the main contactor (M) is closed, and starting current is sent through the resistor (R1-R2) to the motor.

When coil (M) is energized, it closes contacts (M) and provides itself with a sealing circuit from L1 to the stop button. In addition, interlock (M) is opened when coil (M) is energized. As a result, coil (A) is deenergized. Due to the induced current in the copper sleeve around the iron core of A, there is a time delay before the spring closes contact (A) and opens interlock (A). When contacts (A) close, resistor (R1-R2) is shunted out, the motor is connected across the full line voltage, and the motor acceleration is accomplished.

STUDY/DISCUSSION QUESTIONS

1. Describe the manner in which time delay is achieved in a magnetic time limit controller.

2. What adjustment can be made so that the motor will accelerate faster?

Unit 44 Voltage Drop Acceleration

OBJECTIVES

After studying this unit, the student will be able to

- Identify terminal markings for direct-current compound motors and voltage drop acceleration controllers.
- Describe the operation of a voltage drop acceleration controller.
- Connect and troubleshoot compound motors and controllers for voltage drop acceleration.

Voltage drop acceleration is obtainable by using a dc controller which has double coil lockout contactors. These contactors use the voltage drop across the starting resistors to furnish current to the holding coils to obtain time delay.

The starting current of a motor is high. Thus, the voltages across the starting resistors are high and the voltage across the armature is low. However, as the motor accelerates, the voltage drop across the resistors decreases. In the control diagram shown in figure 44-1, the voltage drop across the resistors (R1-R2-R3-R4) is used to send current through the lockout coils (1HA, 2HA, and 3HA) and also to start the time delay. When coil (M) is energized from the start button, contact (M) closes and a heavy starting current passes through R1-R2-R3-R4, the series field, and the motor armature to the line (L2).

The high starting current through the starting resistance produces a large voltage drop across each section of the starting resistance. As a result, a large current is supplied to the holdout coils (1HA, 2HA, and 3HA) and the accelerating contactors are held open.

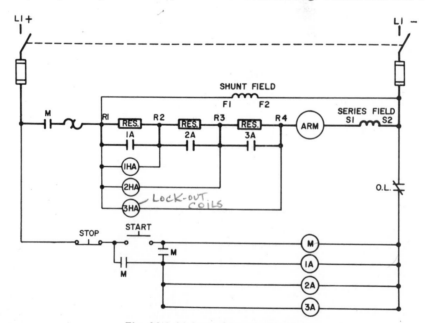

Fig. 44-1 Voltage drop acceleration.

As the motor accelerates and the current through R1-R2 is reduced in value, the current in the holdout coil (1HA) is no longer sufficient to lock out contactor (1A). When contactor (1A) closes, R1-R2 is cut out of the circuit.

As R1-R2 is cut out, the current increases in value. The voltage drop across R2-R3 is high enough to keep contactor (2A) from closing. As the motor continues to accelerate, the current decreases until the voltage drop across R2-R3 is too small to hold contactor (2A) open against the pull of coil (2A). When contactor (2A) closes, it cuts out R2-R3 and the current increases slightly.

The voltage drop across R3-R4 now sends a high current through holdout coil (3HA). Coil (3HA) continues to hold contactor (3A) open against the pull of coil (3A) until the motor accelerates more and the current is reduced. Coil (3HA) can no longer keep contactor (3A) open; thus, contactor (3A) closes and cuts out the last step of resistance R3-R4. By connecting 1HA across R1-R2, 2HA across R1-R3, and 3HA across R1-R4, the operation of the contactors in the sequence 1A, 2A, and then 3A is guaranteed and all of the other contactors can not close simultaneously.

STUDY/DISCUSSION QUESTIONS

1. Are electrically-operated interlocks necessary in the voltage drop acceleration controller? Why?

2. Why is the voltage across R1-R2 high when the motor is started and why does it decrease as the motor accelerates?

Unit 45 Series Relay Acceleration

OBJECTIVES

After studying this unit, the student will be able to

- Identify terminal markings for direct-current compound motors and series relay acceleration controllers.

- Describe the sequence of operation of a dc series relay acceleration controller.

- Connect and troubleshoot dc compound motors and controllers for series relay acceleration.

Dc series relays consist of a few turns of heavy wire and are extremely fast in operation. A spring opens the contacts when the current decreases to a value below that for which the relay is set.

Figure 45-1 shows the schematic diagram of a dc motor controller that provides series relay acceleration. The series relays (1S, 2S, and 3S) are connected to carry the main motor current. To adjust the point at which the relay opens, the spring tension is changed.

When the start button is pressed, a circuit is completed through coil (M). The main contacts (M) close to supply current to the coil of series relay (1S) which, in turn, opens contact (1S). The auxiliary contacts (M) then close to place a shunt around the start button to keep coil (M) energized and the main contacts (M) closed. If this step is not taken, the motor will stop when the start button is released.

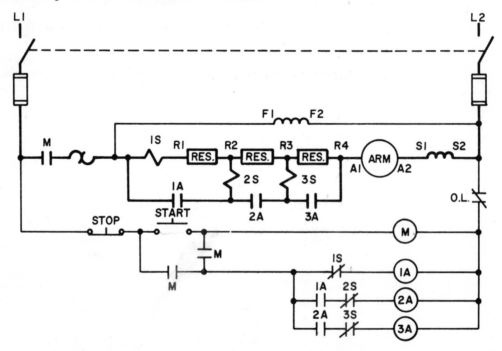

Fig. 45-1 Line diagram of dc series relay acceleration.

The time interval between the closing of the main contacts and the auxiliary contacts on M insures that coil (1A) is not energized before 1S is opened.

The current decreases because of the increase in the speed and the counter emf. As a result, the series relay (1S) can no longer hold its armature open against the tension of the closing spring and contact (1S) closes. When 1S closes, coil (1A) is energized and contact (1A) cuts out the first step of resistance (R1-R2). The current increases to a new value and now has a path from L1 through M, contacts (1A), coil (2S), R2-R3-R4, the armature, and the series field to L2.

As soon as the current reaches its new high value it causes contact (2S) of the series relay (2S) to open before the auxiliary contact (1A) of relay (1A) has time to energize coil (2A).

Again, as the motor speed accelerates and the motor current is reduced, the series relay (2S) can no longer hold contact (2S) open; therefore, coil (2A) is energized. Contact (2A) closes and cuts out the starting resistor (R2-R3). With R2-R3 removed, the current increases to a new high value. This high current through 3S opens contact (3S) before 3A can be energized. Later, as the motor speed increases and the current is reduced, 3S closes and 3A is energized and cuts out the last step of resistance (R3-R4). The motor is now placed directly across the line and accelerates to a normal speed based on the load attached to the motor.

STUDY/DISCUSSION QUESTIONS

1. Explain the difference between a series relay and a shunt relay.
   SERIES RELAY FEW TURNS HEAVY WIRE, FAST. SHUNT RELAY KEEPS MOTOR FROM STOPPING
2. Why is there an increase in armature current with the elimination of each step of resistance?

3. The relay coils (1A, 2A, and 3A) do not become energized when the start button is pressed. Why? CONTACT IS OPEN SO 1A CANNOT BE ENERGIZED

4. What causes the time delay between the closing of the contactors (1A, 2A, and 3A)?

Unit 46 Series Lockout Relay Acceleration

OBJECTIVES

After studying this unit, the student will be able to

- Identify terminal markings for direct-current compound motors and series lockout relay acceleration controllers.

- Describe the sequence of operation of a series lockout relay acceleration controller.

- Connect and troubleshoot dc compound motors and controllers for series lockout relay acceleration.

The series lockout coil (HA) holds the contactor open during the time interval that the starting current of the motor remains high.   The series pull-in coil closes the contactor.

Figure 46-1 is the schematic diagram of a controller which uses series lockout relays to produce the required time delay during motor acceleration. Coils (1A) and (1HA) are the pull in and lock out coils respectively. These coils act on the common contactor (1A) as stated previously in unit 6.

When the start button of the controller shown in figure 46-1 is pressed, coil (M) is energized and main contactor (M) provides a current path from L1 through the main contacts (M), the motor armature, the series field, 1A, 1HA, R1-R2-R3-R4, and 3HA to L2.

The current rises to a high value when the motor starts. This large current through lockout coil (1HA) produces a stronger magnetic pull on contactor (1A) than does coil (1A) which is trying to close 1A. As the motor accelerates, the current is reduced in value. Eventually, a value of current through 1HA is reached that does not permit 1HA to resist the magnetic pull of 1A; thus, contactor (1A) closes.

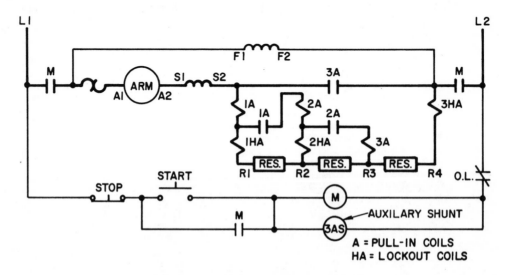

Fig. 46-1 Dc series lockout relay acceleration.

The current path is now from L1 through **M**, the armature, the series field, coil (1A), contact (1A), coils (2A and 2HA), R2-R3-R4, and coil (3HA) to L2. When 1A cuts out resistor R1-R2, the current rises to a high value. This large current through lockout coil (2HA) produces more magnetic pull than does coil (2A), and so contact (2A) is held open. As the motor continues to accelerate, the motor current drops to a lower value and lockout coil (2HA) can no longer resist the pull of 2A, with the result that contactor (2A) closes.

The series resistance R1-R2-R3 is now cut out and the current rises to a higher value. This current through 3HA holds contact (3A) open against the pull of coil (3A) which is trying to close it.   Finally, when the current is reduced in value, 3A closes and all of the resistors and relays are shunted out.

Although it is reasonable to expect that this action will cause all of the relay contacts to drop out, they do not because coil (3AS), which is an auxiliary shunt winding acting on the contactor (3A), is strong enough to hold contact (3A) closed after it is pulled into contact. Coil (3AS) alone is not strong enough to make 3A close its contacts without the aid of coil (3A); however, coil (3AS) is strong enough to hold the contact after it closes.

STUDY/DISCUSSION QUESTIONS

1. All contactors do not close at the same time to shunt out all of the resistors simultaneously. Why? *Because current must be reduced in value first*

2. What are the positions of all of the other contactors when 3A is closed? *They do not drop out.*

3. What is the purpose of coil (3AS)? *Auxiliary shunt winding which acts on contactor (3A) to keep it closed after it's pulled into contact*

Unit 47 Dashpot Motor Control

OBJECTIVES

After studying this unit, the student will be able to

- Describe how a dashpot is used to provide time limit acceleration.
- Describe the use and operating sequence of a field accelerating relay.
- Describe the use and operating sequence of a field failure relay.
- State the uses of multicircuit, time limit dashpot acceleration.
- Connect and troubleshoot dc compound motors and controllers with dashpot, time limit acceleration.

INDIVIDUAL DASHPOT MOTOR CONTROL

Unit 6 stated that the motion of a piston may be retarded by forcing air, oil, or other fluid from one end of a cylinder to the other through a small hole in the piston. This combination of cylinder and piston is called a *dashpot*. The addition of a dashpot to a relay results in a slowing down of the action of the relay.

A dashpot provides a definite timed interval which is used to control acceleration while a motor is being brought up to speed. Resistors inserted in the line ahead of the motor control the acceleration steps. The definite time limit acceleration provided by the dashpots means that the resistors are allowed to stay in the line for a fixed amount of time only before the motor is connected across the line.

Time limit acceleration is generally accepted as the method best suited for a majority of applications. The disadvantage of the compensating time, counter emf, and current limiting methods of acceleration is that the motor may not accelerate sufficiently to close the running contactor that bypasses the resistors. In other words, the resistors will remain in the circuit and may burn out. This hazard is eliminated by the use of time limit acceleration since the running contactor is closed automatically after a definite time interval.

FIELD ACCELERATING RELAY

A field accelerating relay (also known as a full field relay) is used with starters to control adjustable speed motors. This type of relay provides the full field during the starting period and also limits the armature current during sudden speed changes. The coil of the field accelerating relay is connected in series with the motor armature. When there is a current inrush at startup, or when a sudden speed increase causes excessive armature currents, the coil closes the relay contacts. The closed contacts bypass the shunt field rheostat. As a result, the full shunt field strength is provided on starting and excessive armature currents are prevented during speed changes.

FIELD FAILURE RELAY

A field failure relay is a single-pole control relay. The contacts of the relay are connected in the control circuit of the main contactor, and the coil is connected in series with the shunt field. If the shunt field fails, the relay coil is deenergized, and the relay contact opens the circuit to the main line contactor with the result that the motor is disconnected from the line. The field failure relay is a safety feature that can be added to a starter to prevent the motor from speeding up greatly in the event that the field is opened accidentally.

Fig. 47-1 Dc field accelerating relay (left) and dc field failure relay. (Courtesy General Electric Co.)

Referring to figure 47-2, the closing of the start button energizes coil (M), which, in turn, instantaneously closes the heavy power contacts (M) and starts the motor through the starting resistor. The shunt field and field failure relay (FF) are energized at this same moment. The control contacts (FF and M) close and maintain the circuit. The series coil of the field accelerating relay (FA) receives the maximum current flow and closes contacts (FA) to shunt the circuit across the shunt field rheostat. Since the maximum field strength is present, there is more counter emf and excessive armature currents are prevented.

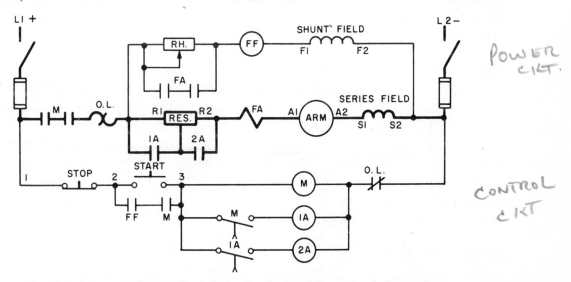

Fig. 47-2 Schematic diagram for definite time limit starting, using dashpot relays, field accelerating relays, and field failure relays.

Fig. 47-3 Dc reduced voltage controller used on steel mill furnace for roof swing and tilt. Controller panel has two reversing controllers, instantaneous overload protection, dynamic braking, armature shunt control, motor brake contactors, and field discharge resistors. Auxiliary ac relays for associated functions are mounted in the base. Panel is 90 inches high and 20 inches deep. (Courtesy General Electric Co.)

Main contactor (M) also activates a normally open (N.O.), delay-in-closing contact (M). At the end of the time delay of this dashpot, it energizes the first accelerating relay (1A). Power contact (1A) cuts out one step of resistance and also activates the dashpot timing cycle (1A). The N.O., delay-in-closing contact (1A) energizes the second accelerating relay (2A) which places the armature across the line.

MULTICIRCUIT, TIME LIMIT DASHPOT ACCELERATION

Multicircuit, time limit controllers are used with series, shunt, or compound-wound dc motors to drive pumps, fans, compressors, and other classes of machinery where the load varies over a wide range and automatic or remote control is desired. Multicircuit controllers use a multifingered contact to cut out the starting resistor in the armature circuit, step by step, as the motor accelerates.

Figure 47-4 shows the control circuit for a time limit dashpot acceleration controller; figure 47-5 shows a typical starter.

The starting resistors used in this type of time limit controller generally are rated for one 10-second start out of each 80 seconds. These resistors are not suitable for continued operation at reduced speeds.

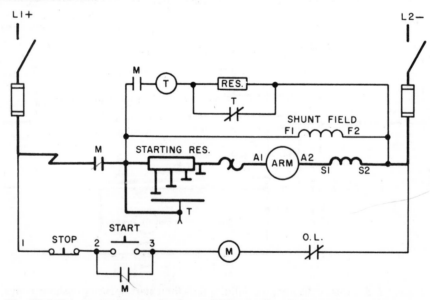

Fig. 47-4 Schematic diagram for dc, multicircuit, time limit dashpot acceleration.

Time limit accelerating controllers using solenoid-operated air or oil dashpots perform both the timing and the starting resistor, short-circuiting functions.

The multifingered contact assembly may be adjusted to meet the conditions under which the motor is expected to operate. Once adjusted, the starter will time accurately the accelerating period regardless of the load conditions and without attention from the operator.

Fig. 47-5 Dc time limit acceleration starter. (Courtesy Square D Co.)

STUDY/DISCUSSION QUESTIONS

1. What protection is available from excessive starting currents in the event the timing relays cut out the starting resistors too rapidly for the connected load?

2. How does a field accelerating relay prevent the excessive current due to a speed change resulting from the adjustment of the rheostat?

3. How do individual dashpot timers, which have only light, single-pole control contacts, control the acceleration steps of a large dc motor?

4. How may more steps of acceleration be added to a starter?

5. Why is use made of double-break contacts or contacts connected in series, such as contacts (FA) across the rheostat and M in the armature circuit in figure 47-2?

6. In figure 47-4, what activates the multifingered dashpot?

7. What does the symbol between L1+ and the mainline contact (M) represent in figure 47-4?

8. Why is a resistor installed in the coil (T) circuit in figure 47-4?

9. Referring to figure 47-4, where is the field failure relay connected and where are its control contacts connected?

10. Can an accelerating relay be added to the starter circuit in figure 47-4? Why?

Unit 48 Pilot Motor-Driven Timer Controller

OBJECTIVES

After studying this unit, the student will be able to

- Identify terminal markings for dc compound motors used with pilot motor-driven timer controllers.

- Describe the sequence of operation of a dc pilot motor-driven timer controller.

- Connect and troubleshoot dc compound motors and controllers with pilot motor-driven timers.

When large ac or dc motors are to be started infrequently, it is possible to cut out steps of starting resistance in succession with the use of a set of contacts on a common shaft driven by a small pilot motor.

The operation of a pilot motor-driven timer controller is shown by the schematic given in figure 48-1. When the start button is pressed, coil (M) energizes the main contactor (M) and the main motor starts. Simultaneously, the normally open contact (M) is closed and the pilot motor causes the moving contacts on the drum controller to start rotating. The moving contact fingers (4, 5, and 6) are successively farther apart from the stationary contacts (4, 5, and 6). As the pilot motor drives the contact fingers, moving contact (4) comes into contact with stationary contact (4) and coil (1A) is energized. When coil (1A) is energized, contacts (1A) close and the first step of the starting resistance (R1-R2) is shunted out. At later intervals, coils (2A) and (3A) are energized in succession, and contacts (2A) and (3A) close to shunt out resistors (R2-R3) and (R3-R4) respectively.

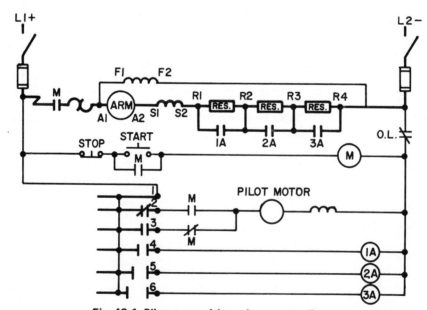

Fig. 48-1 Pilot motor-driven timer controller.

After the entire starting resistance is removed from the circuit, the left end of moving contact (2) leaves stationary contact (2) and the pilot motor stops. Moving contact (3) is touching stationary contact (3), but the pilot motor circuit remains open since the normally closed contact (M) is held open by coil (M).

When the stop button is pressed, coil (M) is deenergized, the normally closed contact (M) energizes the pilot motor, and the drum controller continues to revolve until it is interrupted by the disengaging action of moving contact (3) from stationary contact (3). The controller is now in its original position and is ready to complete another start-stop cycle.

By providing a number of contact fingers and, therefore, a large number of starting resistors, motors ranging in size from 10 h.p. to 100 h.p. may be controlled by the same timer.

If small pilot motors are used with the timer, they will begin to rotate on the first contact points. If larger pilot motors are used, they will not start until a sufficient number of points are cut in to reduce the starting resistance until there is enough current to start the motor. This procedure will not harm either the motor or controller.

STUDY/DISCUSSION QUESTIONS

1. Explain why both the normally open contacts (M) and the normally closed contacts (M) are needed in the pilot circuit.

2. How are the pilot motor contacts positioned for a new start after the stop button is pressed?

Unit 49 Capacitor Timing Starter

OBJECTIVES

After studying this unit, the student will be able to

- Describe the principle of capacitor timing.

- Explain how the capacitor timing principle is applied to an electromagnetic starter.

- Connect and troubleshoot dc compound motors and capacitor timing starters.

For many years, electromagnetic controls have used the capacitor timing principle. That is, if a charged capacitor is discharged through a contactor coil, the contacts of the coil can be held in for a period of time. The length of this time period depends on the following factors: the initial charge on the capacitor, the spring setting of the contactor, the gap between the armature and the core when the armature is "up," and the amount of resistance in the circuit.

Figure 49-1 is the schematic diagram for a starter which contains a capacitor to obtain time delay. The capacitor is charged through a protective resistance (3-4) and the normally closed interlock (M) to make a circuit from 1 to 3. Coil (A) is energized through the interlock (1-5) so that the normally closed contacts (A) open and the resistance (R1-R2-R3) is placed in series with the motor.

When the start button is pressed, coil (M) is energized. The main contacts (M) are closed and the motor is started with full resistance in the circuit. Coil (M) opens the interlocks (1-3) and (1-5) to permit the capacitor to discharge through coil (A) and resistor (3-5). Resistor (3-5) prevents the capacitor from discharging too rapidly.

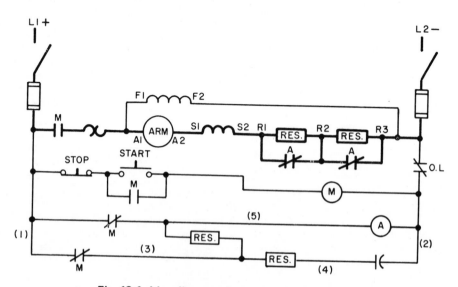

Fig. 49-1 Line diagram of capacitor timing starter.

166

After a period of time, the current in coil (A) is reduced to the point that one armature is released with the result that one step of the starting resistance is cut out of the circuit. As the current continues to decrease, it reaches the value at which the other armature is released. Thus, the other contact (A) shunts out the remainder of the starting resistance.

If the line voltage is slightly below normal, then the motor generally will require more time for startup. However, the lower charge on the capacitor in this situation results in a lower current in coil (A). This means that the armature is released sooner and the acceleration time actually is reduced. Since the acceleration should be kept the same as that for full line voltage, or even increased slightly, a compensating coil is wound on the contactor with coil (A) to keep the time delay independent of the line voltage.

STUDY/DISCUSSION QUESTIONS

1. How can the starter be deenergized completely?

2. In the circuit in figure 49-1, which of the resistors is the capacitor charging protective resistor?

3. If the timing cycle is to be converted so that it is adjustable, what electrical modification is necessary?

4. How are the armatures adjusted to close independently on contactor (A)?

Section 10 Methods of Deceleration

Unit 50 Jogging (Inching) Control Circuits

OBJECTIVES

After studying this unit, the student will be able to

- Define the process of jogging (inching).
- State the purpose of jogging controllers.
- Describe the operation of a jogging control using a control relay.
- Describe the operation of a jogging control using a control relay on a reversing starter.
- Describe the operation of a jogging control using a selector switch.
- Connect and troubleshoot jogging controllers and circuits.

Jogging, or *inching,* is defined by the National Electrical Manufacturer's Association as the quickly repeated closure of a circuit to start a motor from rest for the purpose of accomplishing small movements of the driven machine. The term jogging is often used when referring to across-the-line starters while the term inching is used to refer to reduced voltage starters. In general, however, the terms are used interchangeably because they both prevent a holding circuit.

The control circuits covered in this unit are representative of the various methods that are used to obtain jogging.

Figure 50-2 is a line diagram of a very simple jogging control circuit. The stop button is held open mechanically as shown in figure 50-1. Because the stop button is held open, maintaining contact (M) is unable to hold the coil energized after the start

Fig. 50-1 Mechanical lockout-on-stop pushbutton. (Courtesy Square D Co.)

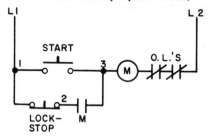

Fig. 50-2 Lock-stop pushbutton in jogging circuit.

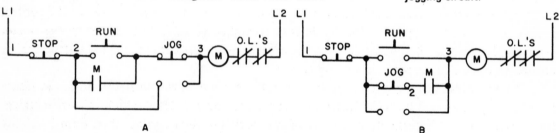

A   B

Fig. 50-3 Line diagrams of simple jogging control circuits.

168

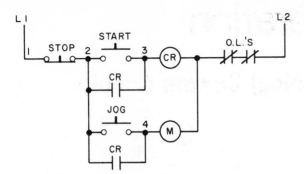

Fig. 50-4 Jogging is achieved with added use of control relay.

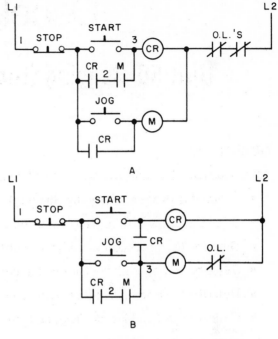

Fig. 50-5 Line diagrams using control relays in typical installations.

button is closed. The disadvantage of a circuit connected in this manner is that the lock-stop safety feature is lost. This circuit can be mistaken for a conventional three-wire control circuit.

Figure 50-3 illustrates other simple schemes for jogging circuits. The normally closed pushbutton contacts on the jog button in figure 50-3B are connected in series with the holding circuit contact on the magnetic starter. When the jog button is pressed, the normally open contacts energize the starter magnet, while the normally closed contacts disconnect the holding circuit. When the button is released, therefore, the starter immediately opens to disconnect the motor from the line. A jogging attachment can be used to prevent the reclosing of the normally closed contacts of the jog button. This device assures that the starter holding circuit is not re-established if the jog button is released too rapidly. Jogging can be repeated by reclosing the jog button and can be continued until the jogging attachment is removed.

Jogging Using a Control Relay

The use of a jogging circuit means that the starter can be energized only as long as the jog button is depressed. As a result, the machine operator has instantaneous control of the motor drive.

The addition of a control relay to a jogging circuit results in even greater control of the motor drive. A control relay jogging circuit is shown in figure 50-4. When the start button is pressed, the control relay is energized and a holding circuit is formed, both for the control relay and the starter magnet. The jog button is connected to form a circuit to the starter magnet. This circuit is independent of the control relay. As a result, the jog button can be pressed to obtain the jogging or inching action.

Other typical jogging circuits using control relays are shown in figure 50-5. In figure 50-5A, the pressing of the start button energizes the control relay, which in turn energizes the starter coil. The normally open starter interlock and relay contact then form a holding circuit around the start button. When the jog button is pressed, the starter coil is energized

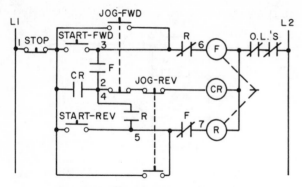

Fig. 50-6 Jogging using control relay
on a reversing starter.

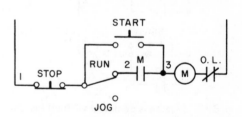

Fig. 50-7 Jogging using a standard duty,
two-position selector switch.

independently of the relay and a holding circuit does not form. As a result, a jogging action can be obtained.

Jogging Using a Control Relay on a Reversing Starter

The control circuit shown in figure 50-6 permits the motor to be jogged in either the forward or the reverse direction while the motor is at standstill or is rotating in either direction. Pressing either the start-forward button or the start-reverse button causes the corresponding starter coil to be energized. The coil then closes the circuit to the control relay which picks up and completes the holding circuit around the start button. While the relay is energized, either the forward or the reverse contactor will also remain energized. If either jog button is pressed, the relay is deenergized and the closed contactor is released. Continued pressing of either jog button results in a jogging action in the desired direction.

Jogging Using a Selector Switch

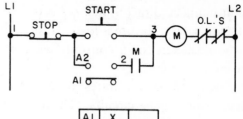

A1	X	
A2		X
	JOG	RUN

Fig. 50-8 Jogging using selector switch —
jog with start button.

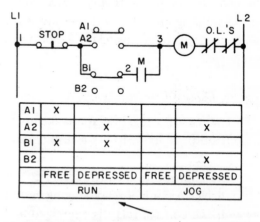

A1	X			
A2		X		X
B1	X	X		
B2				X
	FREE	DEPRESSED	FREE	DEPRESSED
	RUN		JOG	

Fig. 50-9 Jogging using selector pushbutton.

The use of a selector switch in the control circuit to obtain jogging requires a three-element control station with start and stop controls and a selector switch. A standard duty, two-position selector switch is shown connected in the circuit in figure 50-7. The starter maintaining circuit is disconnected when the selector switch is placed in the jog position. The motor is then inched with the start button. Figure 50-8 is the same circuit as that shown in figure 50-7 with the one exception that the selector switch is a heavy-duty, two-position type.

The use of a selector pushbutton to obtain jogging is shown in figure 50-9. In the jog position, the holding circuit is broken, and jogging is accomplished by depressing the pushbutton.

STUDY/DISCUSSION QUESTIONS

1. Why is jogging (inching) included in the section on "Methods of Deceleration"?

2. What is the safety feature of a lock-stop pushbutton?

3. In figure 50-3A, what will happen if both the run and jog pushbuttons are closed?

4. What will happen (refer to figure 50-4) if the start and jog pushbuttons are pushed at the same time?

5. In figure 50-6, (reverse jogging), what will happen if both jog pushbuttons are pushed momentarily?

6. Draw an elementary control diagram of a reversing starter. Use a standard duty selector switch with forward, reverse, and stop pushbuttons with three methods of interlocking.

Unit 51 Plugging

OBJECTIVES

After studying this unit, the student will be able to

- Define what is meant by the plugging of a motor.
- Describe how a control circuit using a zero-speed switch operates to stop (plug) a motor.
- Describe the action of a time-delay relay in a plugging circuit.
- Describe briefly the action of the several alternate circuits which use the zero-speed (plugging) switch.
- Connect and troubleshoot plugging control circuits.

Plugging is defined by NEMA as a system of braking in which the motor connections are reversed so that the motor develops a counter torque which acts as a retarding force. Both rapid stop and quick reversal of motor rotation can be achieved by using plugging controls.

Motor connections can be reversed while the motor is running unless the control circuits are designed specifically to prevent this type of connection. Any standard reversing controller can be plugged, either manually or with electromagnetic controls. Before the plugging operation is attempted, however, there are several factors which must be considered. Included in these factors is the need to determine if methods of limiting the maximum permissible currents are necessary, especially with repeated operations and dc motors. In addition, the driven machine must be examined to insure that repeated plugging will not damage the machine.

PLUGGING SWITCHES

Plugging switches (also called zero-speed switches) are designed to be added to control circuits as pilot devices to provide quick, automatic stopping of machines driven by squirrel cage motors. If the switches are adjusted properly, they will prevent the reversal of the direction of rotation of the controlled drive after it reaches a standstill following the reversal of the motor connections. Several common uses of plugging switches are for machine tools which must stop suddenly at some point in their cycle of operation to prevent inaccuracies in the work or damage to the machine, and for processes in which the machine must stop completely before the next step of work begins. In the latter example, the reduced stopping time means that more time can be applied to production to achieve a greater total output.

Typical plugging switches are shown in figure 51-1. The shaft of a plugging switch is connected mechanically to the motor shaft or to a shaft on the driven machine. The rotating motion of the motor is transmitted to the plugging switch contacts either by a centrifugal mechanism or by a magnetic induction arrangement (eddy current disc) within the switch. The switch contacts are wired to the reversing starter which controls the motor. The switch acts as a link between the motor and the reversing starter, and allows the starter to apply just enough power in the reverse direction to bring the motor to a quick stop.

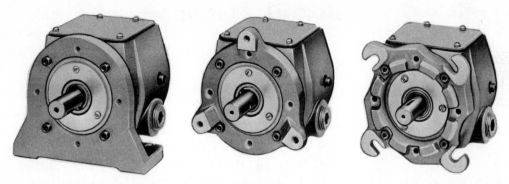

Fig. 51-1 Plugging (zero-speed) switches. Note mounting methods. (Courtesy Allen-Bradley Co.)

Plugging a Motor to a Stop from One Direction Only

In figure 51-2, the forward rotation of the motor closes the normally open plugging switch contact. When the stop button is pushed, the forward contactor drops out, and the reverse contactor is energized through the plugging switch and the normally closed forward interlock. As a result, the motor connections are reversed and the motor is braked

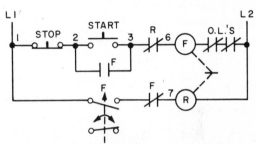

Fig. 51-2 Plugging motor to stop from one direction only.

to a stop. When the motor is stopped, the plugging switch opens to disconnect the reverse contactor. This contactor is used only to stop the motor using the plugging operation and is not used to run the motor in reverse.

Adjustment

The torque that operates the plugging switch contacts varies according to the speed of the motor. To insure that the contacts open and close at the proper time regardless of the motor speed, an adjustable contact spring is used to oppose the torque. To operate the contacts, the motor must produce a torque that will overcome the spring pressure. The spring adjustment is generally made with screws that are readily accessible when the switch cover is removed.

Installation

To obtain the greatest possible accuracy in braking, the switch should be driven from the shaft with the highest available speed that is within the operating speed range of the switch.

The plugging switch may be driven by gears, by a chain drive, or by use of a direct flexible coupling. The latter method, a direct flexible coupling connected to a suitable shaft on the driven machine, is the preferred way of driving the switch. The coupling must be flexible since the centerline of the motor or machine shaft and the centerline of the plugging switch shaft are difficult to align accurately enough to use a rigid coupling. The switch must be driven by a positive means and so a belt drive should not be used. In addition, a positive drive must be used between the various parts of the machine being controlled, especially where these parts have large amounts of inertia.

The starter used for this type of circuit is a reversing starter that interchanges two of the three-phase motor leads for a three-phase motor, reverses the direction of current through the armature for a dc motor, and reverses the relationship of the running and starting windings for a single-phase motor.

Motor Rotation

Experience shows that there is no way of predetermining the direction of rotation of motors when the phases are connected externally in the proper sequence. This factor is an important consideration for the electrician and the electrical contractor when the applicable electrical code or specifications require that each phase wire of a distribution system be color coded.

If, when looking at the shaft end of a motor, it runs counterclockwise rather than in the desired clockwise direction, then the electrician must reconnect the motor leads at the motor. For example, assume that many three-phase motors are to be connected and the direction of rotation of all the motors must be the same. If counterclockwise rotation is desired, when looking at the shaft end of the motor, then the supply phase should be connected to the motor terminals in the proper sequence, T_1, T_2, and T_3. If the motor does not rotate in the desired counterclockwise direction using these connections, the leads may be interchanged at the motor. Once the proper direction of rotation is established, the remaining motors can be connected in a similar manner if they are from the same manufacturer. If the motors are from different manufacturers, they may rotate in different directions even when all the connections are similar and the supply lines have been phased out for the proper phase sequence and color coded. The process of correcting the rotation may be difficult if the motors are located in a nearly inaccessible place.

Lockout Relay

The zero-speed switch can be equipped with a lockout relay or a safety latch relay. A relay of this type provides a mechanical means of preventing the switch contacts from closing unless the motor starting circuit is energized. This safety feature insures that if the motor shaft is turned accidentally, the plugging switch contacts do not close and start the motor. The relay coil generally is connected to the T1 and T2 terminals of the motor.

PLUGGING WITH THE USE OF A TIMING RELAY

A time-delay relay may be used in a motor plugging circuit, figure 51-3. Unlike the zero-speed switch, however, this control scheme does not compensate for a change in the load conditions on the motor. The circuit shown in figure 51-3 is satisfactory for a constant load condition once the timer is preset. If the emergency stop button is pushed momentarily and the normally open circuit is not completed, then the motor will coast to a standstill. (This action is true of the normal stop button also.) But, if the emergency stop button is pushed to complete the normally open circuit of the pushbutton, contactor (S) is energized through the closed contacts (TD and R). Contactor (S) closes and reconnects the motor leads, thus causing a reverse torque to be applied. When the relay coil is deenergized, the opening of contact (TD) can be retarded. The time lag is set so that contact (TD) opens at or near zero motor shaft rpm.

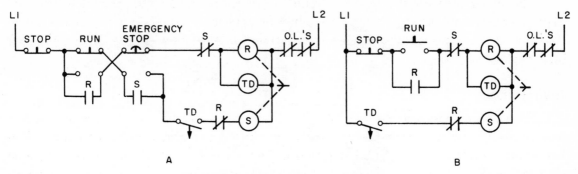

Fig. 51-3 Plugging with time-delay relay.

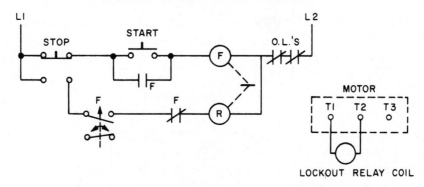

Fig. 51-4 Holding stop button stops motor in one direction.

ALTERNATE CIRCUITS FOR PLUGGING SWITCH

The circuit in figure 51-4 is for operation in one direction. When the stop pushbutton is pressed and immediately released, the motor and the driven machine coast to a standstill. If the stop button is held down, then the motor is plugged to a stop.

In figure 51-5, the motor may be started in either direction, and when the stop button is pressed, the motor can be plugged to a stop from either direction.

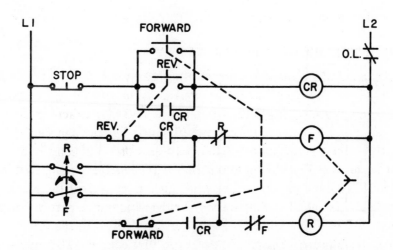

Fig. 51-5 Pressing stop button stops motor in either direction.

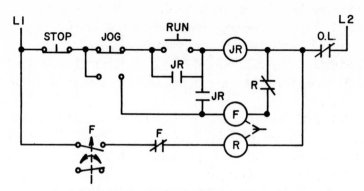

Fig. 51-6 Pressing stop button stops motor in one direction.

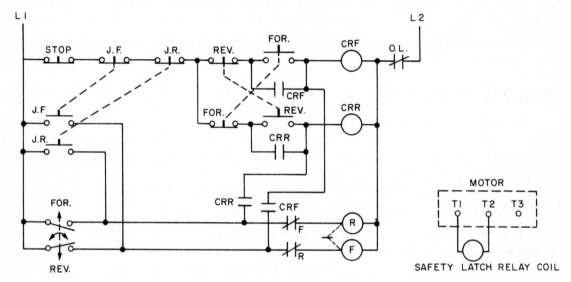

Fig. 51-7 Use of control jogging relays stops motor in either direction.

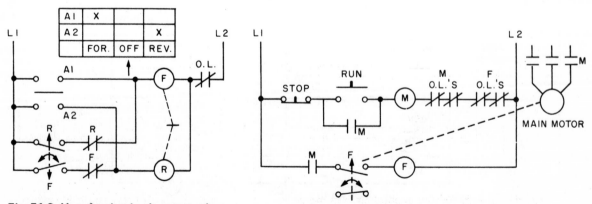

Fig. 51-8 Use of maintained contact selector switch.

Fig. 51-9 Use of plugging switch as speed interlock.

Operation is in one direction in figure 51-6. The motor is plugged to a stop when the stop button is pressed. Jogging is possible with the use of a control relay.

Figure 51-7 shows a circuit which can control the direction of rotation of a motor in either direction. Jogging in either the forward or reverse direction is possible if control jogging relays are used. The motor can be plugged to a stop from either direction by pressing the stop button.

The circuit in figure 51-8 provides control in either direction using a maintained contact selector switch with forward, off, and reverse positions. The plugging action is available from either direction of rotation when the switch is turned to the off position. Low-voltage protection is not provided with this circuit.

The circuit in figure 51-9 allows motor operation in one direction. The plugging switch is used as a speed interlock. The solenoid, or the coil (F), will not operate until the main motor reaches its running speed. A typical application of this circuit is to provide an interlock for a conveyor system. The feeder conveyor motor cannot be started until the main conveyor is in operation.

ANTIPLUGGING PROTECTION

Antiplugging protection, as defined by NEMA, is obtained if a device prevents the application of a counter torque until the motor speed is reduced to an acceptable value. An antiplugging circuit is shown in figure 51-10. With the motor operating in one direction, a contact on the antiplugging switch opens the control circuit of the contactor used to achieve rotation in the opposite direction. This contact will not close until the motor speed is reduced, after which the other contactor can be energized.

Alternate Antiplugging Circuits

The direction of rotation of the motor is controlled by the motor starter selector switch as shown in figure 51-11. The antiplugging switch completes the reverse circuit only when the motor slows to a safe, preset speed. Undervoltage protection is not available.

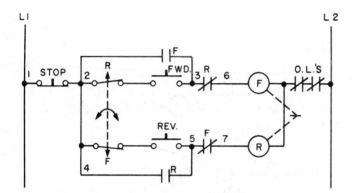

Fig. 51-10 Antiplugging protection; the motor is to be reversed but not plugged.

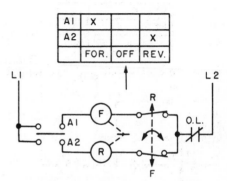

Fig. 51-11 Antiplugging with rotation direction selector switch.

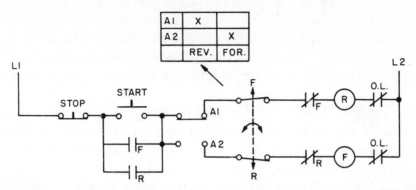

Fig. 51-12 Antiplugging circuit using a selector switch and providing low-voltage protection.

In figure 51-12, the direction of rotation of the motor is selected using the maintained contact, two-position selector switch. The motor is started with the pushbutton. The direction of rotation cannot be reversed until the motor slows to a safe, preset speed. Low-voltage protection is provided by a three-wire, start-stop, pushbutton station.

STUDY/DISCUSSION QUESTIONS

1. In figure 51-2, what is the purpose of N.C. contact (F)?
2. Can a time-delay relay be used satisfactorily in a plugging circuit? Explain.
3. In what position must a plugging switch be mounted?   Explain.
4. What is the preferred method of connecting the plugging switch to the motor or the driven machine?
5. What happens if the zero-speed switch contacts are adjusted to open too late?
6. What is the purpose of the lockout relay or safety latch relay?
7. What happens if the reverse pushbutton is closed when the motor is running in the forward direction, as in figure 51-5?
8. What alternate methods of stopping are provided by the circuit in figure 51-4?
9. If the motor in figure 51-6 is plugging to a stop and the operator suddenly wants it to inch ahead or run, what must he do?
10. Is it necessary to push the stop button when changing the direction of rotation in figure 51-7? Explain.
11. In figure 51-9, how is the feeder motor protected from an overload?
12. In figure 51-8, what happens to the motor running in the forward direction if the power supply is lost for 10 minutes?
13. What is antiplugging protection?
14. During normal operation, when do the antiplugging switch contacts close?
15. If the supply lines are in proper phase sequence (L1, L2, and L3) and are connected to their proper terminals on the motor, T1, T2, and T3, will all the motors rotate in the same direction? Why?

Unit 52 Electric Brakes

OBJECTIVES

After studying this unit, the student will be able to

- Describe the general operation of electric (magnetic) brakes, both ac operated and dc operated.
- State the advantages of dc shunt-wound brakes over ac-operated brakes.
- Describe the function of typical braking circuits for hoists and cranes.
- Connect and troubleshoot motors and braking controllers.

The increased use of electricity has brought about a need for more powerful motors running at increased speeds. These motors frequently require a faster means of stopping than is obtained merely by disconnecting the power from the motor. When machines require one or more motors and more precise control, and when there is less time available to wait for the machines to coast to a standstill, then the necessity to provide faster stopping methods becomes even greater.

Electric brakes, also called *magnetic brakes, friction brakes* and *mechanical brakes,* have been in use since the early 1900s. An electric brake generally consists of two friction surfaces, or shoes, which can be made to bear on a wheel on the motor shaft. Spring tension holds the shoes on the wheel and braking occurs as a result of the friction between the shoes and the wheel. A solenoid mechanism is used to release the shoes.

In a magnetically operated brake, the shoes are held in a released position by a magnet as long as the magnet coil is energized. However, if a pilot device interrupts the power or there is a power failure, then the brake shoes are applied instantly to provide a fast, positive stop.

The coil leads of an alternating-current magnetic brake normally are connected directly to the motor terminals. If a reduced-voltage starting scheme is used, the brake coil should be connected to receive full voltage.

Fig. 52-1 Solenoid brake used on machine tools, conveyors, small hoists, and similar devices (Courtesy Cutler-Hammer, Inc.)

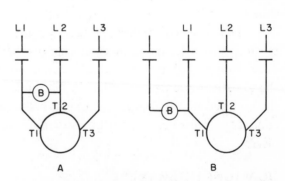

Fig. 52-2 Ac brake coil connections for across-the-line starting. Control relays and contactors also are used to control magnetic brakes.

Fig. 52-3 Torque motor-operated brake for
ac service on cranes, hoists, and elevators.
(Courtesy Cutler-Hammer, Inc.)

Fig. 52-4 Solenoid-operated disc brake for
mounting on motor end bell.
(Courtesy Cutler-Hammer, Inc.)

Magnetic brakes provide a smooth braking action which makes them particularly adaptable to high-inertia loads. Since these brakes apply and remove the braking pressure smoothly in either direction, they are often used on cranes, hoists, elevators, and other machinery where it is desirable to limit the shock of braking. A brake suitable for such service is shown in figure 52-3.

The brake in figure 52-3 is operated by a vertically-mounted torque motor. When power is applied to the brake torque motor, the shaft of the motor turns and lifts the operating lever of a ball jack to release the brake. When the brake is fully released, the torque motor is stalled across the line. When power is released from the brake, a heavy compression spring counteracts the jack lever so that the brake is applied quickly.

MAGNETIC DISC BRAKES

In general, disc brakes can be installed wherever shoe brakes are used. In addition, disc brakes can be installed where considerations of appearance and space prohibit the use of shoe brakes.

Magnetic disc brakes are used on machine tools, cranes, hoists, conveyors, printing presses, saw mills, index mechanisms, overhead doors, and other installations. The control of the torque and wear of disc brakes is approximately the same as that for shoe brakes in that the adjustments are very similar. A disc brake is a self-enclosed unit that is bolted directly to the end bell of the motor. (This end bell operates on the motor shaft.) The braking action consists of pressure released by a solenoid and applied by a spring on the sides of a disc or discs.

DC MAGNETIC BRAKES

Dc magnetic brakes are constructed in a manner similar to that of ac shoe brakes. However, for dc magnetic brakes, the operating coils are shunt wound for across-the-line connections and series wound for connecting in series with dc motor armatures.

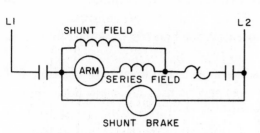

Fig. 52-5 Dc shunt brake connection with motor connections

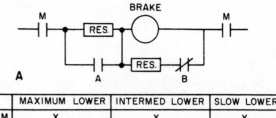

	MAXIMUM LOWER	INTERMED LOWER	SLOW LOWER	OFF
M	X	X	X	
A	X			
B			X	X

B

Fig. 52-6 Varying friction braking.

Shunt-wound brakes pick up (release) at approximately 80 percent of the rated voltage. Series-wound brakes pick up (release) at about 40 percent of the full load current of the motor and drop out (set) when the current reaches 10 percent of the full load value.

Fig. 52-7 Series brake connected with series motor.

The brake linings for dc magnetic brakes are similar to those for ac brakes. The linings are made of molded or woven asbestos with high friction resistance and are bonded or riveted to steel backplates. Brakes range in size from approximately one inch to 30 inches in diameter. Large diameter shunt-wound brakes are available with a relay and protective resistor to permit operation at high speed. Brakes with smaller diameters pick up (release) and drop out (set) quickly without this added feature. The large diameter, high-speed brakes release quickly; when the armature closes, the protective resistor is inserted in the circuit. This action reduces the holding current to a low value and keeps the coil cooler, resulting in the quick setting of the brake. High-speed brakes are suitable for any duty. They can be operated on alternating current if a rectifier unit is added.

The operation of dc magnetic brakes tends to be somewhat sluggish due to the induction present if the manufacturer has not attempted to overcome it to speed the operation of the brake. The use of dc brakes for shunt-wound operation has several advantages over ac brakes, including:

- Laminated magnets or plungers are not used.

- Magnet and armature are cast of steel.

- No destructive hammer blow

- Fast release and fast set

- No ac chatter

- Short armature movement

- No coil burnout due to shoe wear which affects the air gap.

Figure 52-5 shows the connection of a dc brake in a circuit which does not allow armature regeneration through the brake coil. As a result, the brake is prevented from setting.

Figure 52-6A shows a shunt brake control circuit (this is a small portion of a crane hoist elementary diagram). The target table in figure 52-6B indicates the contacts which are closed for different crane lowering operations. Note how the brake is weakened by the use of resistors in series and in parellel.

Series-wound brakes generally are used with series motors and must have the same full-load current rating as the motor. Series-wound brakes may be used for heavy hoist applications and for many metal processing applications, such as steel mill drives and conveyors.

STUDY/DISCUSSION QUESTIONS

1. Why are magnetic brakes used?

2. How do electric brakes operate?

3. When an ac reduced-voltage starter is used, how should the electric brakes be wired for release?

4. Why should protective resistors be provided on coils for high-speed shunt brakes?

5. In figure 52-6, which contacts are closed when the brake coil receives full voltage?

6. What precautions must be taken when matching a series-wound brake to a motor?

7. How are electric brakes adjusted in general?

8. What may cause a deenergized brake to slip under load with maximum spring tension?

Unit 53 Dynamic Braking

OBJECTIVES

After studying this unit, the student will be able to

- Describe what is meant by the process of dynamic braking.
- Describe the general series of controller connections that can be made to obtain the hoisting, lowering, and braking operations at different speeds.
- List several advantages of dynamic braking.
- Describe three methods of providing dynamic braking for light-duty equipment.
- Explain how dynamic braking can be provided for a synchronous motor.
- Connect and troubleshoot motors and dynamic braking controllers.

A motor can be stopped by disconnecting it from the power source. However, a faster means of stopping a motor is achieved if the motor is reconnected so that it acts as a generator. This method of braking is called *dynamic braking*.

When the motor is reconnected so that the field is excited and there is a low resistance path across the armature, the generator action converts some of the mechanical energy of rotation to electrical energy (as heat in the resistors). The result is that the motor slows down sooner. However, as the motor slows down, the generator action decreases, the current decreases, and the braking lessens. As a result, a motor cannot be stopped by dynamic braking alone.

A motor driving a crane hoist will let the load down too quickly if it is simply disconnected from the line. Some kind of braking is essential. A slight variation of the dynamic braking method described above works very well for this situation.

A series motor is used because of its characteristics for lifting and moving heavy loads. A solenoid brake connected in the line keeps the motor shaft from turning when there is no line current. To stop the hoist, therefore, it is necessary only to open the circuit to the motor.

To lower a load, the motor is reconnected as a shunt generator. That is, the series field is connected in series with a resistor and then the combination is connected directly across the line. Other possible connection schemes for series motors may be used.

The armature may be connected to a resistance to obtain dynamic braking as described above; however, it is more satisfactory to connect the armature to the line as well. As a result, some of the mechanical energy is put back in the line and the resistors are not required to handle so much heat. This process is called *regenerative braking*.

If the control is moved from the hoisting or lowering positions to the stop position, the load should be slowed by dynamic braking before the mechanical brake takes hold to reduce the wear on the mechanical brake bands.

The schematic diagrams in figure 53-1 show how the controller connections can be changed to obtain the hoisting, lowering, and braking operations at different speeds. To simplify these diagrams, the control contactors and other components are not shown.

In figure 53-1, note that for maximum hoist operation (a), the dc series motor and the brake receive the full line voltage and the maximum current.

In the intermediate hoist position (b), the motor is slowed when a resistor is added in series with the motor. The resistor is not large enough to allow the brake to set or drag, however.

In the slow hoist position (c), some current is bypassed around the motor by the parallel resistor. As a result, the motor is slowed and the brake receives enough current to remain open.

In the lower positions, (d)-(f), regeneration back into the supply system is possible. The dissipating resistor regulates the field and the brake.

The brake is deenergized in the off position with maximum braking pressure applied.

Dynamic braking is a simple and safe method of emergency or safety braking. Since the motor is converted to a self-excited generator to provide the slowing action, the braking does not depend on an outside source of power. Machines operating at higher speeds have braking problems not found with lower operating speeds. The need for a more rapid initial slowdown is essential if accidents are to be avoided. Dynamic braking is applied in about one-fifth of the time required to set the majority of shunt brakes.

Due to space limitations, some drives can not be equipped with electric brakes. In other

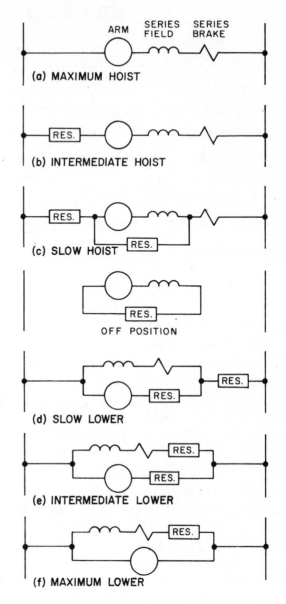

Fig. 53-1 Reconnecting a crane controller for different operations.

cases, the inertia of the brake wheel is objectionable, since it retards acceleration and deceleration.

An advantage of dynamic braking is shown on outside cranes or ore bridges where wet or icy rails cause the wheels to slip during stopping. When the wheels slip, the dynamic braking decreases, the wheels begin rotating again, and the crane or bridge is stopped in a shorter distance than is possible with locked wheels.

On high-speed drives, even when brakes are used, it is recommended that dynamic braking be provided to reduce the drive speed to a low value before the brakes are set, as in the case of an ore bridge trolley where severe braking may endanger the operator.

Dynamic braking can be initiated by track-type limit switches installed in the end zones

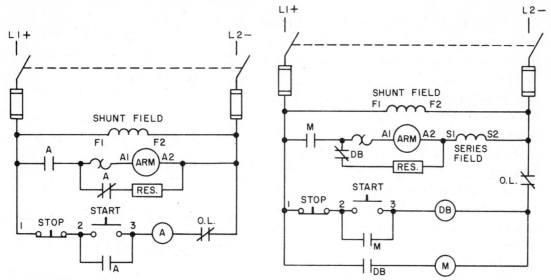

Fig. 53-2 Dynamic braking connections on motor starter.

Fig. 53-3 Dc motor starter modified with dynamic braking.

of overhead traveling cranes. With dynamic braking provided in this manner, cranes and other equipment can be stopped quickly and automatically when there is a power failure or the overload relays trip, regardless of the position of the master control handle. The braking torque is equally effective in both directions of travel.

DYNAMIC BRAKING FOR LIGHT-DUTY EQUIPMENT

There are many different methods of providing dynamic braking for small production equipment. In figure 53-2, the reduced-voltage starting mechanisms are eliminated to simplify the diagram. When the stop button is depressed, normally closed contact (A) completes the braking circuit through the braking resistor. Note that the shunt field must be energized for deceleration as well as acceleration; full field strength must be available for both. If a rheostat is used in the shunt field for speed control, the resistance is cut out manually or automatically. The disconnect switch should be open when the machine is not in service.

Figure 53-3 is a modification of figure 53-2. The circuit in figure 53-3 insures that the braking resistor is not connected across the line. If the series field is used in the braking circuit, it is necessary to determine if the dc compound motor characteristic is differential or cumulative. If it is a cumulative compound motor, then the series field current direction must be reversed when it is used in the braking circuit. Automatic operation can be provided for such an action.

Figure 53-4A shows a diagram for a braking circuit which uses the starting resistance for braking. This circuit may be satisfactory for a number of situations if the starting and stopping cycle does not exceed the capacity of the resistors. A speed regulator is designed for continuous-duty operation. The fixed resistor (R) in the braking circuit is a current-limiting resistor. It may not be required when a speed regulator is used because of the increased capacity. Figure 53-4B diagrams the control circuit for adjusting the braking speed. Resistor (R) is a current-limiting resistor which prevents a short circuit on the armature. The value of R is selected so maximum braking speed is obtained when all resistance is cut out by the speed regulator.

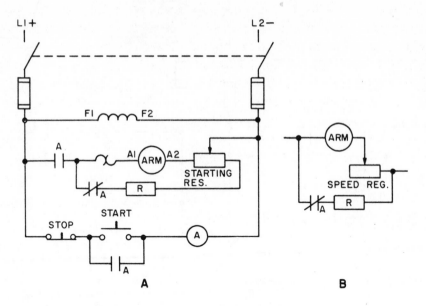

Fig. 53-4 Using the starting resistance for braking.

DYNAMIC BRAKING FOR A SYNCHRONOUS MOTOR

Because of the similarity in the construction of a synchronous motor and an alternator (ac generator), a synchronous motor can be reconnected as an alternator to provide faster stopping. The kinetic energy of the rotor and the driven machine is converted to electrical energy by generator action and then to heat by the dissipating resistors. A method developed by the author for the dynamic braking of a synchronous motor is illustrated in the elementary diagram of figure 53-5. When the start button is pressed, contactor (A) is energized and opens the resistor circuits connected to the motor leads in the control panel. Contactor (A) energizes coil (M) to maintain the circuit and start the motor across the line. Time-delay relay coil (TD) is energized and its timing cycle begins. After the motor is brought up to speed by the windings in the rotor, the N.O., delay-in-closing contact (TD) energizes the dc contactor (B) to supply current to the field. The field discharge

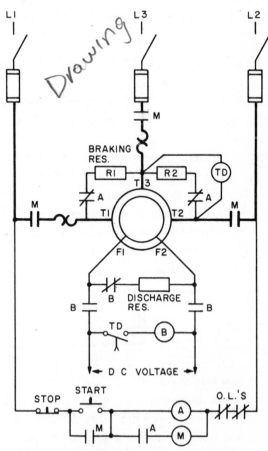

Fig. 53-5 Dynamic braking applied to synchronous motor.

resistor circuit is opened as well. When the stop button is pressed, the ac supply is removed from the stator, but the timer coil remains energized so that its contact remains closed. As a result, direct current continues to flow through the rotor. As the rotation continues, the magnetic lines of force of the rotor cut across the stator windings to generate a current which keeps the timing relay contact (TD) closed. The contact remains closed as long as the rotor maintains a voltage drop across the braking resistor (R2). Direct current is removed automatically from the rotor when it is almost at a standstill. This timed, semiautomatic method of synchronizing was selected to illustrate this method of braking for reasons of simplicity. With a simple modification, it is readily adaptable to any completely automatic, synchronizing control system.

STUDY/DISCUSSION QUESTIONS

1. After dynamic braking occurs, why is a mechanical brake generally necessary?

2. What becomes of the energy used to brake the motor dynamically for the two methods described?

3. Referring to figure 53-1, what will happen if the controller is placed in the off position from a maximum lower position?

4. The brake in figure 53-1 is not shown in the off position. Why?

5. Why is it necessary to open the disconnect switch when the machine is not in operation (figure 53-2)?

6. Why may it be necessary to reverse the polarity of the series field of a compound motor if it is used in the dynamic braking circuit?

7. Referring to figure 53-4B, why is resistor (R) necessary?

8. In figure 53-4A, why is resistor (R) necessary?

9. Draw a schematic diagram of a circuit to reverse the polarity of the series field of a compound motor used in the automatic braking circuit when the stop button is pressed (see figure 53-3 and question 6 above).

10. Referring to figure 53-5, what keeps the time-delay coil (TD) energized after coil (M) is deenergized?

Unit 54 Electric Braking

OBJECTIVES

After studying this unit, the student will be able to

- Describe the method of operation of a typical electric braking controller.
- List advantages of the use of an electric brake on an ac induction motor.
- Connect and troubleshoot electric braking controllers.

APPLICATION

An efficient and effective brake for a standard drive, ac squirrel cage motor is provided by an electric braking controller. This type of controller is designed to stop motors driving high-inertia loads such as roving or spinning frames. Electric brakes also may be used with conveyors, machine tools, woodworking machines, textile machinery, rubber mills, and processing machinery.

Electric braking controllers can be conveniently installed on new or existing machines and have several advantages over conventional brakes. For example, it is sometimes difficult to install conventional electromechanical brakes on existing equipment without extensive rebuilding. In addition, mechanical brakes require considerable mechanical maintenance, adjustment, and periodic parts replacement due to worn parts.

OPERATION

The principle of electric braking is applied to standard ac squirrel cage or wound-rotor motors. When such a motor is to be stopped, direct current is applied to one or all of the three phases of the motor after the ac voltage is removed. As a result, the motor is braked quickly and smoothly to a standstill. An electric braking controller provides a smoother positive stop because the braking torque decreases rapidly as the speed approaches zero. This braking torque generally can be adjusted by the use of tapped resistors.

When compared to the method of single-phase braking that is commonly used, three-phase braking improves the efficiency of stopping and reduces the heat buildup in the motor.

One brake stop is equivalent to a normal start; therefore, the allowable number of starts that can be made without overheating the motor must be reduced accordingly.

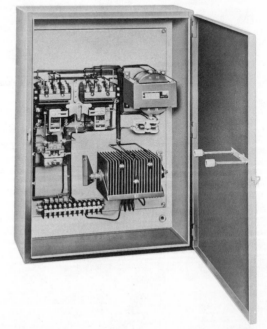

Fig. 54-1 Electric braking control for starting, protecting, and braking an ac induction motor. (Courtesy Square D Co.)

CONSTRUCTION

A self-contained transformer and rectifier provide the direct current required for braking. An easily adjustable timer is used to provide dependable timing of the braking cycle. A standard reversing starter can be used to supply power to the motor: ac during normal operation and dc during braking.

Terminal blocks are available for motor and control connections. Taps on current-limiting resistors mean that adjustments of the braking torque can be made over a wide range. Controllers consisting of braking units only can be obtained for use with existing motor starters.

SEQUENCE OF OPERATION OF ELECTRIC BRAKING CIRCUIT

Figure 54-2 shows the schematic diagram of a typical electric braking circuit. When motor starter (M) is energized, the motor operates as soon as the start button closes. The normally closed contact of the start button insures that braking contactor (B) is deenergized before the motor is connected to the ac supply.

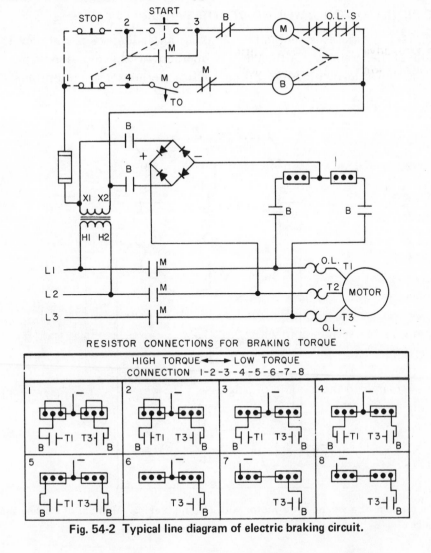

Fig. 54-2 Typical line diagram of electric braking circuit.

While M is energized, its normally closed interlock contact is open and its normally open time-delay contact is closed. When the stop button is pressed, therefore, M drops out; when interlock (M) recloses, braking contactor (B) is energized. B remains energized until the time delay contact (M) times open. This timing contact is adjustable and should be set so that contactor (B) drops out as soon as the motor comes to a complete stop.

Braking contactor (B) connects all three motor leads to a source of dc through the rectifier and transformer. Direct current applied to an induction motor polarizes it to a stationary magnetic field and causes it to brake to a stop. The braking torque and the speed of braking can be varied by reconnecting the tapped resistor as shown in figure 54-2, page 189.

Dc and ac supplies must not be connected to a motor at the same time. Therefore, it is very important that the motor starter and braking contactor have adequate interlocks. It is recommended that the motor starter always be included as an integral unit with the brake. If a braking unit only must be applied to a separate motor starter, then mechanical interlocking is sacrificed. In this case, the start button must be equipped with a normally closed contact and the motor starter must have extra normally closed interlock contacts.

ELECTRIC BRAKING FOR A WOUND-ROTOR MOTOR

Figure 54-3 illustrates a method of braking an induction motor developed under the direction of the author. Although any method of starting is possible, across-the-line starting is shown in figure 54-3. Speed control methods also are optional for wound-

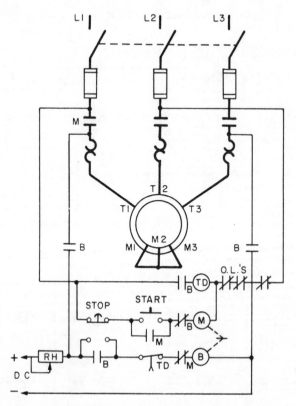

Fig. 54-3 Line diagram of a wound-rotor motor with electric braking. Methods of starting and speed control are optional.

rotor motors. The circuit in the figure shows that the secondary resistors are eliminated and the slip rings are connected together.

The motor is started by pressing the start button to energize starter (M). The motor may be brought to a coasting stop by pressing the stop button slightly. If a fast stop is required, the stop button must be depressed completely to make contact across the normally open portion of the heavy-duty pushbutton. As a result, contactor (B) is energized through the N.C., delay-in-opening timed contact and closed interlock (M). Ac is removed from the stator by contacts (M), and dc is applied by contacts (B) to establish a "stationary" magnetic field. A rapid deceleration of the rotor results.

Timing contact (TD) is adjustable and should be set to drop out contactor (B) as soon as the rotor comes to a complete stop. The braking torque and the speed of braking can be varied by adjusting the applied dc and the rotor secondary resistors. Braking voltage is kept to approximately 10 percent of the motor nameplate voltage.

The braking speeds of motors depend on the motor ratings and the attached load. For example, a one-horsepower motor with a heavy inertia load, such as a flywheel or conveyor, can be stopped easily in about one second or less, if necessary. A 125-horsepower, slow-speed motor can be stopped, if necessary, in about two or three seconds.

It is questionable whether such sudden stops harm the motor or the driven equipment. The effect of stopping by this method is much less pronounced than that of a mechanical friction brake. The braking action of an electric brake is cushioned and resilient, somewhat like moving a block of steel with a rubber band. The maximum braking effect occurs at about 15 to 20 percent of the rated motor speed.

STUDY/DISCUSSION QUESTIONS

1. Why does three-phase braking reduce motor heating as compared to single-phase braking?

2. Differentiate between the ac and dc magnetic fields applied to an induction motor stator.

3. Electric braking can be applied to what type of ac motor?

4. Why is electric braking smooth and resilient?

5. Braking current is passed through overload heaters; however, the overload control circuit contacts are in the ac starting circuit. How does this protect the motor?

6. How often can a motor be stopped using electric braking?

7. There is no danger of the motor reversing after it is braked to a stop. Why?

Section 11 Motor Drives

Unit 55 Direct Drives and Pulley Drives

OBJECTIVES

After studying this unit, the student will be able to

- State the advantages of direct and pulley drives.
- Install directly-coupled motor drives and pulley motor drives.
- Check the alignment of the motor and machine shafts, both visually and with a dial indicator.
- Install motors and machines in the proper positions for maximum efficiency.
- Calculate pulley sizes using the equation:

$$\frac{\text{Drive rpm}}{\text{Driven rpm}} = \frac{\text{Driven Pulley Diameter}}{\text{Drive Pulley Diameter}}$$

[handwritten annotations: "motor" above Drive, "Load" below Driven, "Load" above Driven Pulley Diameter, "Motor" below Drive Pulley Diameter]

DIRECTLY COUPLED DRIVE INSTALLATION

The most popular choice of running speed and the most economical speed for an electric motor is about 1800 rpm. However, the majority of electrically-driven constant speed machines operates at speeds below 1800 rpm. As a result, these machines must be provided with either a high-speed motor and some form of mechanical speed reducer or a low-speed, directly coupled motor.

Synchronous motors can be adapted for direct coupling to machines operating at speeds from 3600 rpm to about 80 rpm, with horsepower ratings ranging from 20 to 5,000 and above. A rough rule of thumb suggested by the Electric Machinery Manufacturing Co. is that synchronous motors are less expensive to install than squirrel cage motors if the rating exceeds one h.p. per rpm. This rule, however, considers only the first cost and does not take into account the higher efficiency and better power factor of the synchronous motor. When the motor speed matches the machine input shaft speed, a simple mechanical coupling is used, preferably a flexible coupling.

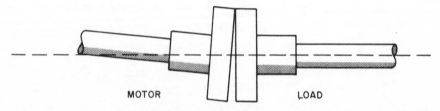

MOTOR   LOAD

Fig. 55-1 Every alignment check must be made from positions ninety degrees apart, or at right angles to each other.

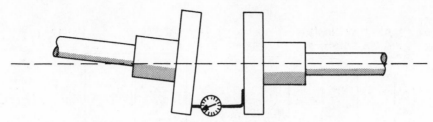

Fig. 55-2 Angular check of direct motor couplings.

Trouble-free operation can be obtained by following several basic recommendations from the Allis Chalmers Manufacturing Co. for the installation of directly coupled drives and pulley or chain drives. When connecting the motor to its load, the alignment of the devices must be checked more than once from positions at right angles to each other. For example, when viewed from the side, two shafts may seem to be exactly in line, but when viewed from the top, figure 55-1, it is evident that the motor shaft is cocked at an angle. A dial indicator, figure 55-2, should be used to check the alignment of the motor and the driven machinery, or, if a dial indicator is not available, a feeler gauge can be used.

Another important check during installation is to make sure the shafts of the motor and the driven machine are not bent. Both the machine and the motor should be rotated together, just as they rotate when the machine is running, and then rechecked for alignment. After the angle of the shafts is aligned, the shafts may appear to share the same axis. However, as shown in figure 55-3, the axes of the motor and the driven machine may be off center. When viewed again from a position ninety degrees away from the original position, it can be seen that the shafts are not on the same axis.

To complete the alignment of the devices, the motor should be moved until the rotation of both shafts shows that they are lying on the same axis when viewed from four positions spaced ninety degrees apart around the shafts. The final test is to check the starting and running currents with the connected load to insure that they do not exceed specifications.

Low-speed, directly coupled induction motors have several disadvantages, however. They usually have a low power factor and low efficiency. Both of these characteristics increase the electric power costs. As a result, induction motors are rarely used for operation at speeds below 500 rpm.

Constant speed motors are available with a variety of speed ratings. In general, the highest possible speed is used to reduce the size, weight, and cost of the motor. At five horsepower, a 1200 rpm motor is almost 50 percent larger than an 1800 rpm motor; at 600 rpm the motor is well over twice as big as the 1800 rpm motor. In the range from 1200 rpm

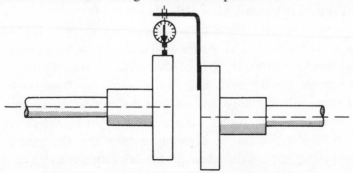

Fig. 55-3 Axis alignment of direct motor couplings.

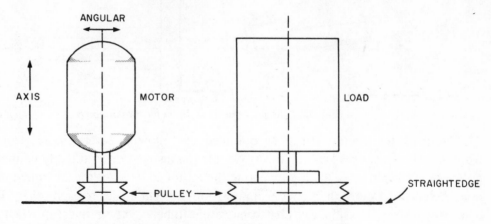

Fig. 55-4 Using straightedge for angular and axis alignment.

to 900 rpm, where the size and cost disadvantages are not an overwhelming factor, low-speed, directly coupled motors can be used. For example, this type of motor is used on most fans, pumps, and compressors.

PULLEY DRIVES

Installation

Flat belts, V-belts, chains, or gears are used on motors so that smooth speed changes at a constant rpm can be achieved. For speeds driven below 900 rpm, it is practical to use an 1800 rpm or 1200 rpm motor connected to the driven machine by a V-belt or a flat belt.

Machine shafts and bearings give long service when the power transmission devices are properly installed according to the manufacturer's instructions.

Offset drives can be lined up more easily than direct drives, regardless of whether V-belts, gears, or chain drives are used. Both the motor and load shafts must be level. A straightedge can be used to insure that the motor is aligned on its axis and that it is at the proper angular position so that the pulley sheaves of the motor and the load are in line, figure 55-4. When belts are installed, they should be tightened just enough to assure a nonslip engagement – no more. The less cross tension there is, the less wear there will be for the bearings involved. Proper and firm positioning and alignment are necessary not only to control the forces that cause vibration but also the forces that cause thrust.

The designer of a driven machine usually determines the motor mount and the type of drive to be used. As a result, the installer has little choice in the motor location. In many flat belt or V-belt applications, however, several choices may be available to the construction wireman or maintenance electrician. If a choice is available, the motor should be placed where the force of gravity helps to increase the grip of the belts. A vertical drive, as in figure 55-5, can cause problems because gravity tends to pull the belts away from the lower sheave. To counteract this, the belts require far more tension than the bearings should have to withstand.

There are also correct and incorrect placements for horizontal drives. The location where the motor is to be installed can be determined from the direction of rotation of the motor shaft. It is recommended that the motor be placed with the direction of rotation such that the belt slack is on top. In this position, the belt tends to wrap around the sheaves.

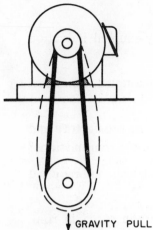

Fig. 55-5 Greater belt tension is required in vertical drives where gravity opposes good belt traction. This greater tension generally exceeds that which the bearings should withstand.

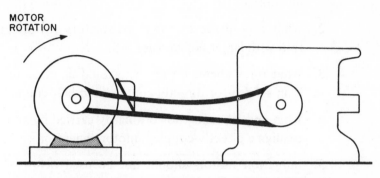

Fig. 55-6 Direction of motor rotation can be used to advantage for good belt traction (with slack at top) if motor is placed in proper position.

This problem is less acute with V-belts than it is with flat belts or chains. Therefore, if rotation is to occur in both directions, V-belts should be used. In addition, the motor should be placed on the side of the most frequent direction of rotation, figure 55-6.

Pulley Speeds

Motors and machines frequently are shipped to installation locations lacking pulleys or with pulleys of incorrect sizes. The drive and driven speeds are given on the motor and machine nameplates or in the descriptive literature accompanying the machines.

Four quantities must be known to set up the machinery with the correct pulley sizes: the drive rpm, the driven rpm, the diameter of the drive pulley, and the diameter of the driven pulley. If three of these quantities are known, then the fourth quantity can be determined. For example, if a motor runs at 3600 rpm, the driven speed is 400 rpm, and there is a four-inch pulley for the motor, then the size of the pulley for the driven load can be determined from the following equation.

$$\frac{\text{Drive rpm}}{\text{Driven rpm}} = \frac{\text{Driven Pulley Diameter}}{\text{Drive Pulley Diameter}}$$

$$\frac{3600}{400} = \frac{X}{4} \quad \text{or} \quad \frac{36\cancel{00}}{4\cancel{00}} = \frac{X}{4}$$

By cross multiplying and then dividing, we arrive at the pulley size required:

$$\begin{array}{r} 36\text{-inch diameter pulley} \\ \hline 4\,)\,\overline{144} \end{array}$$

If both the drive and driven pulleys are lacking, the problem can be solved by estimating a reasonable pulley diameter for either pulley and then using the equation with this value to find the fourth quantity.

STUDY/DISCUSSION QUESTIONS

1. What are the disadvantages of low-speed, directly-coupled induction motors? *Low power factor & Low efficiency*

2. What type of ac motor is satisfactory for directly-coupled use at low rpm with larger horsepower ratings? *Synchronous motor*

3. What three alignment checks should be made to insure satisfactory and long service for directly-coupled and belt-coupled power transmissions?

4. What special tools, not ordinarily carried by an electrician, are required to align a directly-coupled, motor-generator set? *1. Dial Indicator 2. Feeler Gauge*

5. Induction motors should not be used below a certain speed in rpm. What is this speed? *500 R.P.M*

6. What is the primary reason for using a pulley drive? *Long Service Speed Change*

7. How tightly should V-belts be adjusted? *Non-slip engagement*

8. Refer to figure 55-6 and assume the motor is to rotate in the opposite direction. How can the belt slack be maintained on top? *Use V-Belt Change location to opposite side of DRIVEN Device*

9. A machine delivered for installation has a two-inch pulley on the motor and a six-inch pulley on the load. The motor nameplate reads 1800 rpm. At what speed in rpm will the driven machine rotate?

Handwritten margin notes:
Check Positions Right Angles to each other
1. Side 90° vertical horiz.
2. Angular — Longst.
3. Axis Belt Driven

Handwritten work:

$$\frac{Drive\ Rpm}{Driven\ Rpm} \qquad \frac{Driven\ Pulley\ Diameter}{Drive\ Pulley\ Diameter}$$

$$\frac{1800}{200} = \frac{X}{6}$$

$$\frac{X\ 200}{200} = \frac{10800}{200} = 54 \qquad \frac{1800}{X} \qquad \frac{6}{2}$$

$$\frac{1800}{200} = \frac{X}{6} \qquad \frac{6X}{6} = \frac{3600}{6} = 600$$

$$\frac{4\,1800}{10800} = \qquad \frac{2X}{2} = \frac{108}{2}$$

Unit 56 Gear Motors

OBJECTIVES

After studying this unit, the student will be able to

- Describe the basic operation of a gear motor.
- State where gear motors are used generally.
- Describe the three standard gear classifications.
- Select the proper gears for the expected load characteristics.

Many industrial machines require power at slow speeds and high torque. At speeds of 780 rpm and lower, it is customary to use a chain drive, a belt drive, a separate speed reducer coupled to the motor, or a gear motor.

The gear motor is a speed reducing motor that gives a direct power drive from a single unit without the necessity of external gearing and the maintenance such gearing requires. A gear motor usually consists of a standard 1800 rpm ac or dc motor and a sealed gear train correctly engineered for the load. This assembly is mounted on a single base as a one-package, enclosed power drive. The advantage of this unit is its extreme compactness. A gear motor actually is smaller than a low-speed standard motor of the same horsepower.

In the gear motors, the motor-shaft pinion drives the gear or series of gears in an oil bath that is linked with the output shaft. This type of arrangement is usually the most economical and convenient way to obtain low speeds of approximately one rpm to 780 rpm.

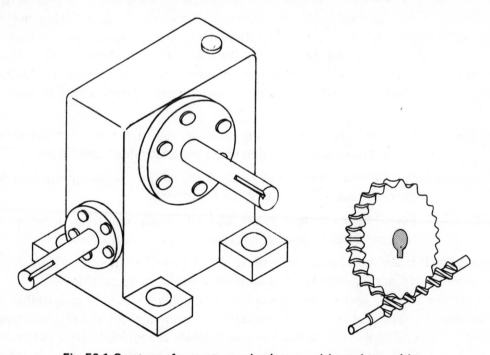

Fig. 56-1 One type of separate speed reducer: a right-angle gear drive.

197

Fig. 56-2 Cutaway of speed reducing gear motor (double reduction).
(Courtesy U. S. Electrical Motors)

One-unit gear motors are available with shafts parallel to each other or at right angles, with polyphase, single-phase, or direct-current voltages, and with horsepower ratings ranging from approximately 1/6 h.p. to 200 h.p.

Gear motors can be selected from any of the many different motor types, running at either constant or adjustable speed. The selection of the control equipment for the motor is the same as with any other motor of the same type.

When selecting a gear motor, an important consideration is the degree of gear service and gear life based on the load conditions to which the motor will be subjected. Gear motors are divided into three classes. Each class uses different gear sizes to handle specific load conditions, and each class gives about the same life for the gears. The American Gear Manufacturer's Association has defined three operating conditions commonly found in industrial service and has established the following three standard gear classifications to meet these conditions.

Class I. For steady loads within the motor rating of 8 hours per day duration, or for intermittent operation under moderate shock conditions.

Class II. For 24-hour operation at steady loads within the motor rating, or 8-hour operation under moderate shock conditions.

Class III. For 24-hour operation under moderate shock conditions, or 8-hour operation under heavy shock conditions.

For conditions that are more severe than those covered by Class III gears, a fluid drive unit may be incorporated in the assembly to cushion the shock to an acceptable value.

To achieve multiple speeds, separate units are available with a shift comparable to that of an automobile. These units must be assembled with the motor and the driven machine. Since the amount of power lost in gearing is very small, the multiple drive has essentially constant horsepower; that is, as the output speed is decreased, the torque is increased.

STUDY/DISCUSSION QUESTIONS

1. Why do many industrial machines use gear motors instead of low rpm induction motors? ~~High~~ Torque   Compact Economical

2. What is the principal difference in gear motor classifications? Constant or adj. speeds
   Shock condition

3. In general, what is the maximum speed in rpm of gear motors? 1800 RPM   output shaft

4. What is the classification of a gear motor that is subjected to continual reversing during an eight-hour production shift? Class III

5. Why is the output shaft on a gear reduction motor larger than the shaft of a standard motor? Obtain low speed with large shaft.
   Greater Torque on output of gear train

Unit 57 Variable Frequency Drives

OBJECTIVES

After studying this unit, the student will be able to

- Describe the operating principle of variable frequency drives.
- State why variable frequency drives are used and the advantages of such drives.

The speed of an ac squirrel cage induction motor depends upon the frequency of the supply current and the number of poles of the motor. Thus, a frequency changer can be used to vary the speed of this type of motor.

To obtain variable frequency, for one or more motors from a fixed frequency power source, an alternator is driven through an adjustable mechanical speed drive. The voltage is regulated automatically during frequency changes. In this system, an ac motor drives a variable cone pulley, or sheave, which is belted to another variable pulley on the output shaft. When the relative diameters of the two pulleys are changed, the relative speed between the input and the output can be controlled.

As the alternator speed is varied, the frequency varies. In situations where more than one motor is involved, such as in conveyor systems, all of the motors are connected electrically to the power system of the alternator. As a result, each motor receives the same frequency and operates at the same speed. If a high-speed motor and a slow-speed gear motor are connected to the same variable frequency circuit, they will change speeds simultaneously and proportionately. It is possible to obtain smooth acceleration of motors attached to the alternator by increasing the frequency. Electric braking using suitable controls can be used in this system.

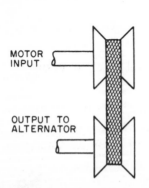

Fig. 57-1 Variable pitch pulleys; mechanical method of obtaining continuously adjustable speed from constant speed shaft.

Fig. 57-2 Variable power unit to control one or more motors at coordinated speeds. (Courtesy U. S. Electrical Motors.)

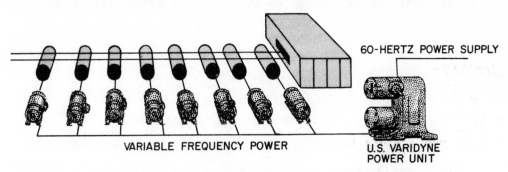

Fig. 57-3 **Coordinating many drive motors, all of standard ac, squirrel cage construction.**

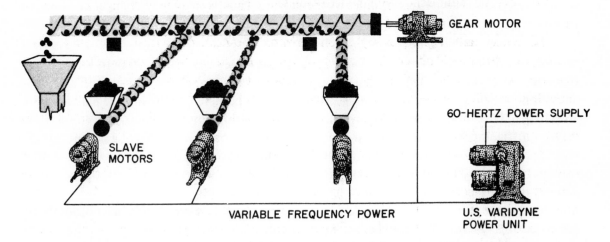

Fig. 57-4 **Screw conveyors may be controlled individually by variable drive *slave motors*. Once the motors are set, they rotate at the same proportional rpm, controlled by the variable frequency power unit supplying them.**

Variable frequency power drives are used on many different types of conveyor systems. For example, automobile assembly lines and coordinated screw conveyors for the proportional mixing of ingredients make use of these power drives.

STUDY/DISCUSSION QUESTIONS

1. What is the formula for finding the speed of an ac squirrel cage induction motor? $S = \dfrac{120 \times f}{P}$

2. If several motors driving an overhead conveyor system are connected to a variable frequency supply line and the frequency is changed from 40 hertz to 80 hertz, what effect does this have on the speed of the motors?   *same freqed same speed*

3. How is the ac frequency controlled in an alternator?   *speed varies freq. "*

4. If many feeder conveyor motors of different poles (speed) are connected to a main conveyor and to a variable frequency power unit, can increased production be achieved by increasing the frequency of the power unit? Explain.   *No same freq. same speed*

5. What is a *slave* motor?   *Extra motor controlled by power unit*

Unit 58 Magnetic Clutch and Magnetic Drive

OBJECTIVES

After studying this unit, the student will be able to

- State several advantages of the use of a clutch in a drive.
- Describe the operating principles of magnetic clutches and drives.
- Distinguish between single and multiple-face clutches.
- Connect and troubleshoot magnetic clutch and magnetic drive controls.

ELECTRICALLY CONTROLLED MAGNETIC CLUTCHES

Machinery clutches were designed originally to engage very large motors to their loads after the motors had reached running speeds. Clutches are used to provide smooth starts for operations in which the material being processed may be damaged by abrupt starts, figure 58-5, page 204. In addition, they are used for starting high inertia loads, since the starting may be difficult for a motor sized to handle the running load. When starting conditions are severe, a clutch inserted between the motor and the load means that the motor can run within its load capacity. The motor will take longer to bring the load up to speed, but the motor will not be damaged.

As more automatic cycling and faster cycling rates are required in industrial production, more electrically controlled clutches are being used.

Fig. 58-1 Magnetic clutches in cement mill service. Note slip rings for clutch supply.

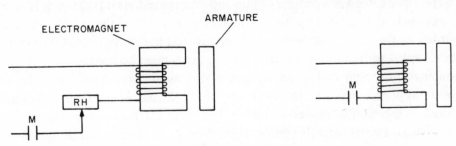

Fig. 58-2 Principle of operation of electrically controlled clutches: (A) gradual clutch engagement; (B) more rapid clutch engagement.

Single-Face Clutch

The single-face clutch consists of two discs: one is the field member and the other is the armature member. The operation of the clutch is similar to that of the electromagnet in a motor starter. When current is applied to the field winding disc through collector (slip) rings, the two discs are drawn together magnetically. The friction face of the field disc is held tightly against the armature disc to provide positive engagement between the rotating drives. When the current is removed, a spring action separates the faces to provide a definite clearance between the discs.

Multiple-Face Clutch

Multiple-face clutches are also available. In a double-face clutch, both the armature and field discs are mounted on a single hub with a double-faced friction lining supported between them. When the magnet of the field member is energized, the armature and field members are drawn together. They grip the lining between them to provide the driving torque. When the magnet is deenergized, a spring separates the two members and they rotate independently of each other.   Double-face clutches are available in sizes up to 78 inches in diameter.

A water-cooled magnetic clutch is available for applications which require a high degree of slippage between the input and output rotating members. Uses for this type of clutch include tension control (wind-up and payoff) and cycling (starting and stopping) operations

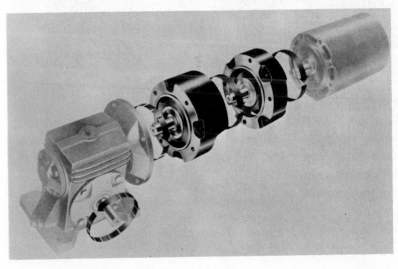

Fig. 58-3 Electric brake and electric clutch modules. Brake is shown on the center left on the driven machine side of the clutch. (Courtesy Warner Electric Brake and Clutch Co.)

in which large differences between the input and the output speeds are required. Flowing water removes the heat generated by the continued slippage within the clutch. A rotary water union mounted in the end of the rotor shaft means that the water cooled clutch can not be directly end-coupled to the prime mover; chains or gears must be used.

A combination clutch and magnetic brake disconnects the load from the drive and simultaneously applies a brake to the load side of the drive. Magnetic clutches and brakes are often used as mechanical power-switching devices in module form. Figure 58-3 shows a drive line with an electric clutch (center right) and an electric brake. Remember that the quicker the start or stop, the shorter the life.

Magnetic clutches are used on automatic machines for four general functions: starting, running, cycling, and torque-limiting. The combinations and variations of these functions are practically limitless.

MAGNETIC DRIVES

The magnetic drive couples the motor to the load magnetically. While the magnetic drive can be used as a clutch, it also is adaptable to an adjustable speed drive.

The electromagnetic (or eddy current) coupling is one of the simpler methods of obtaining an adjustable output speed from the constant input speed of squirrel cage motors.

There is no mechanical contact between the rotating members of the magnetic drive and thus there is no wear. Torque is transmitted between the two rotating units by an electromagnetic reaction created by an energized coil winding. The slip between the motor and load can be controlled continuously, with more precision, and over a wider range than is possible with the mechanical friction clutch.

As shown in figure 58-6, the magnet rotates within the steel ring or drum. There is an air gap between the ring and the magnet. The magnetic flux crosses the air gap and penetrates the iron ring. The rotation of the ring with relation to the magnet generates eddy currents and magnetic fields in the ring. Magnetic interaction between the two units transmits torque from the

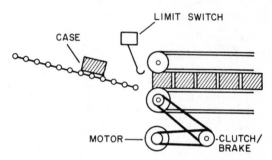

Fig. 58-4 **Case sealer is used to hold top of carton while glue is drying. In this application, cartons come down gravity conveyor and hit switch in front of sealer. Clutch on sealer drive is engaged, moving all cartons in sealer forward. When new carton passes trip switch, brake is engaged and clutch disengaged. Positioning provides even spacing of cartons, insuring that they are in the sealer for as long a time as possible.**

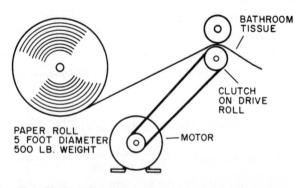

Fig. 58-5 **To prevent tearing, a cushioned start is required on drive roll that winds bathroom tissue off large roll. Roll is five feet in diameter and weighs 500 pounds when full. Pickup of thin tissue must be very gradual to avoid tearing. Application also can be used for filmstrip processing machine.**

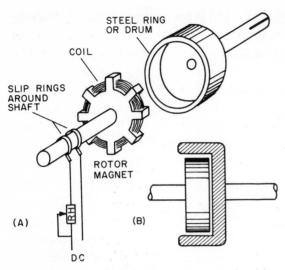

STEEL RING OR DRUM

COIL

SLIP RINGS AROUND SHAFT

ROTOR MAGNET

(A)

(B)

DC

Fig. 58-6   (A) Open view of magnetic drive assembly; (B) spider rotor magnet rotates within ring.

Fig. 58-7 Two magnetic drives driven by 100-h.p., 1200-rpm induction motors mounted on top. Machines are used in typical sewage pumping plant. Pumps are mounted beneath floor of drives. (Courtesy Electric Machinery Mfg. Co.)

motor to the load. This torque is controlled with a rheostat which manually or automatically adjusts the direct current supplied to the electromagnet through the slip rings.

A greater refinement in automatic control to regulate and maintain the output speed is possible when the electromagnetic drive responds to an input or command voltage. The magnetic drive can be used with any type of actuating device or transducer that can provide an electrical signal. For example, electronic controls and sensors which detect liquid level, air and fluid pressure, temperature, and frequency can provide the input required.

A tachometer generator provides feedback speed control in that it generates a voltage proportionate to its speed. Any changes in load condition will change the speed. The resulting generator voltage fluctuations are fed to a control circuit which increases or decreases the magnetic drive field excitation to hold the speed constant.

Where a magnetic drive meets application requirements, it is frequently a desirable choice to provide an adjustable speed. Magnetic drives are used for cranes, hoists, fans, pumps, compressors, and other applications requiring an adjustable speed.

STUDY/DISCUSSION QUESTIONS

1. How is the magnetic clutch engaged and disengaged?

2. What devices may be used to energize the magnetic clutch?

3. Which type of drive is best suited for maintaining large differences in the input and output speeds? Why?

4. What is meant by feedback speed control?

5. How is the magnetic drive used as an adjustable speed drive?

Unit 59 DC Variable Speed Control — Motor Drives

OBJECTIVES

After studying this unit, the student will be able to

- Describe the operating principles of dc variable speed control motor drives.
- State how above and below dc motor base speeds may be obtained.
- List three advantages of dc variable speed motor drives.
- List modifications which can be made to packaged variable speed control drives to meet specific applications.
- Connect and troubleshoot dc variable speed control motor drives.

Adjustable speed drives are available in convenient units that include all necessary control circuits. In general, these packaged variable speed drives operate on ac power and provide a large choice of speeds within given ranges.

Some drive requirements are so exacting that the ac motor drive is not suitable. The dc motor possesses many of the characteristics that an ac motor provides, and in addition has characteristics that cannot be obtained with an ac motor. Because the dc shunt motor with adjustable voltage control is so versatile, it can be adapted to a large variety of applications.

In the integral horsepower range, the motor-generator set is the most widely used method of obtaining variable speed control. The ac service in industrial plants must be converted to dc for use with these controls. By using a system called the Ward Leonard system in which a dc shunt generator with field control is driven by an induction motor, smooth and easily controlled dc is obtained.

Recall from previous instruction that the following factors must be present if electricity is to be generated electromagnetically for heavy power consumption.

- A magnetic field
- A conductor
- Relative motion between the two.

The motion in the dc variable speed drive of figure 59-2 is maintained by the steady driving force of an induction motor. The generator voltage is increased by increasing the magnetic flux in the generator field. The magnetic flux of the generator field is increased by decreasing the resistance of the field rheostat to allow more current to flow through the shunt winding; the greater the current, the stronger the magnetic field.

Fig. 59-1 Packaged-type motor-generator with dc variable speed control system supplied from ac. (Courtesy Square D Co.)

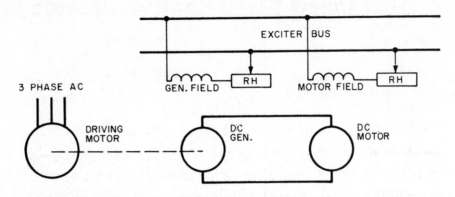

Fig. 59-2 Basic electrical circuit of a dc variable speed control system.

The speed and torque of the system shown in figure 59-2 can be controlled by adjusting the voltage to the field or to the armature or both. Speeds above the motor base speed are obtained by weakening the motor shunt field with a rheostat. Speeds below the motor base speed are obtained by weakening the generator field which lowers the generator voltage supplying the dc armature. The motor should have a full shunt field for speeds lower than the base speed to give the effect of almost continuous control rather than step control of the motor speed.

In figure 59-2, the motor used to furnish power from the line may be a three-phase induction motor as shown, or it may be a dc motor. Once the driving motor is started, it runs continuously at a constant speed to drive the dc generator.

The armature of the generator is coupled electrically to the motor armature. If the field strength of the generator is varied, the voltage from the dc generator can be controlled to send any amount of current to the dc motor. As a result the motor can be made to turn at many different speeds. Because of the inductance of the dc fields and the time required by the generator to build up voltage, extremely smooth acceleration or deceleration is obtained from zero rpm to speeds greater than the base speed.

The field of the dc generator can be reversed manually or automatically with a resulting reversal of the rotation of the motor.

The generator field resistance can be changed manually or automatically by the time-delay relays operated by a counter emf across the motor armature.

Electrically controlled variable speed drives offer a wide choice of speed ranges, horsepower and torque characteristics, means for controlling acceleration and deceleration, and methods of manual or automatic operation. A controlling tachometer feedback signal may be driven by the dc motor driveshaft. This is a refinement of the system to obtain constant speed and depends on the type of application, the speed, and the degree of response required (similar to the magnetic drive automatic control). In addition to speed, the controlling feedback signal may be set to respond to pressure, tension, or some other transducer function.

The circuit most widely used with a motor-generator set to obtain adjustable voltage is shown in figure 59-3, page 208, with exciter closed connections. An ac motor, usually a squirrel cage motor, drives the dc generator and exciter. The exciter may be operated as a self-excited machine and set to give constant potential output.

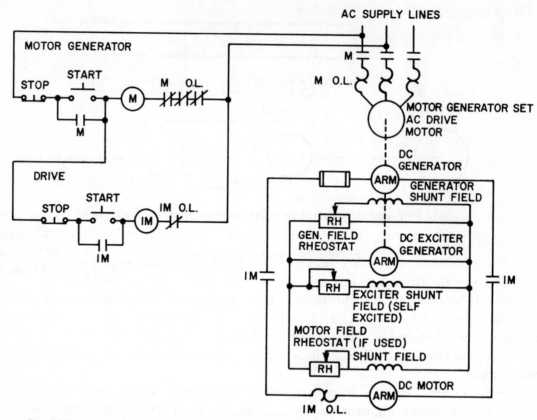

Fig. 59-3 Electrical power and control connections for dc variable speed control system.

The generator field is connected directly across the constant potential exciter. The motor field may be weakened to provide speeds above the base speed after the generator voltage is increased to its maximum value.

The motor-generator set has the one inherent advantage over rectifier-type dc supplies; that is, the ability to regenerate.

Assume that in figure 59-3, the generator rheostat is set at its maximum point and the motor is running at base speed. If the generator voltage is decreased by adjusting the rheostat, the motor countervoltage will be higher than the generator voltage and the current reverses. This gives rise to reverse torque in the machine, and the motor slows down (dynamic braking). When used on metal working, textile and paper processing machines, and for general industry use, this dynamic braking feature is very desirable.

MODIFICATIONS

To meet specific operating requirements, the packaged variable speed control drives can be furnished with a wide variety of modifications, including provisions for:

- Multimotor drive
- Reversing by reversing the generator field
- Reversing by motor armature control
- Dynamic and regenerative braking

- Motor-operated rheostat operation

- Preset speed control

- Jogging at preset reduced speed

- Current or voltage (speed) regulation

- Timed rate of acceleration and deceleration

- Current-limit acceleration and deceleration

The generator field resistance can be changed manually or automatically by time relays operated by a counter emf across the motor armature. Motor-operated rheostats are used in automatic speed control operations also.

The above modifications are often used in combination as required by the specific application. Some control schemes eliminate the standard exciter and provide other methods of energizing the generator and motor fields. Some systems eliminate the motor-generator set entirely and supply the armature and field power directly to the motor from ac lines through controlled rectifiers.

STUDY/DISCUSSION QUESTIONS

1. What is meant by the base speed of a dc motor?

2. How is above-base speed obtained?

3. How is sub-base speed for a dc motor obtained?

4. What advantage does the motor-generator type of variable speed drive have over rectifier type dc supplies?

5. If the ac driving motor of a self-excited motor-generator set is reversed, does this reverse the direction of rotation of the dc motor? Explain.

6. How does reversing the generator polarity reverse the connected motor or motors?

7. Why should the motor have full shunt field for below-base speed operation?

8. To increase generator voltage: (a) should the resistance of the rheostat be increased or decreased; (b) what change in current condition is required in the shunt field? Why?

Appendix-Glossary of Terms

Accelerating Relay. An accelerating relay is any type of relay used to aid in starting a motor or to accelerate a motor from one speed to another. Accelerating relays may function by: (1) motor armature current (current limit acceleration); (2) armature voltage (counter emf acceleration); or (3) definite time (definite time acceleration).

Accessory (control use). A device that controls the operation of magnetic motor control. (Also see Master Switch, Pilot Device, and Pushbutton.)

Across-the-line. A method of motor starting which connects the motor directly to the supply line on starting or running. (Also called Full Voltage Control.)

Alternating Current (Ac). A current changing both in magnitude and direction; most commonly used current.

Ambient Temperature. The temperature surrounding a device.

Ampere. Unit of electrical current.

ASA. American Standards Association.

Automatic. Self-acting, operating by its own mechanism when actuated by some triggering signal, as for example, a change in current strength, pressure, temperature, or mechanical configuration.

Automatic Starter. A self-acting starter which is completely controlled by master or pilot switches or other sensing devices; designed to control automatically the acceleration of a motor during the acceleration period.

Auxiliary Contacts. Contacts of a switching device in addition to the main circuit contacts; auxiliary contacts operate with the movement of the main contacts.

Blowout Coil. Electromagnetic coil used in contactors and starters to deflect an arc when a circuit is interrupted.

Branch Circuit. A branch circuit is that portion of a wiring system extending beyond the final overcurrent device protecting the circuit.

Brake. An electromechanical friction device to stop and hold a load. Generally electric release spring applied — coupled to motor shaft.

Breakdown Torque (of a motor). The maximum torque that a motor will develop with the rated voltage applied at the rated frequency, without an abrupt drop in speed. (ASA)

Busway. A system of enclosed power transmission that is current and voltage rated.

Capacitor-Start Motor. A single-phase induction motor with a main winding arranged for direct connection to the power source and an auxiliary winding connected in series with a capacitor. The capacitor phase is in the circuit only during starting. (NEMA)

Circuit Breaker. Automatic device which opens under abnormal current in carrying circuit; circuit breaker is not damaged on current interruption; device is ampere, volt, and horsepower rated.

Contact. A conducting part which acts with another conducting part to complete or to interrupt a circuit.

Contactor. A device to establish or interrupt an electric power circuit repeatedly.

Controller. A device or group of devices that governs, in a predetermined manner, the delivery of electric power to apparatus connected to it.

Controller Function. Regulate, accelerate, decelerate, start, stop, reverse, or protect devices connected to an electric controller.

Controller Service. Specific application of controller. General Purpose: Standard or usual service. Definite Purpose: Service condition for specific application other than usual.

Current Relay. A relay which functions at a predetermined value of current. A current relay may be either an overcurrent relay or an undercurrent relay.

Dash Pot. A dash pot consists of a piston moving inside a cylinder filled with air, oil, mercury, silicon, or other fluid. Time delay is caused by allowing the air or fluid to escape through a small orifice in the piston. Moving contacts actuated by the piston close the electrical circuit.

Definite Time (or Time Limit). Definite time is a qualifying term indicating that a delay in action is purposely introduced. This delay remains substantially constant regardless of the magnitude of the quantity that causes the action.

Definite-purpose Motor. Any motor designed, listed, and offered in standard ratings with standard operating characteristics or mechanical construction for use under service conditions other than usual or for use on a particular type of application. (NEMA)

Device. A unit of an electrical system that is intended to carry but not utilize electrical energy.

Direct Current (Dc). A continuous nonvarying current in one direction.

Disconnecting Means (Disconnect). A device, or group of devices, or other means whereby the conductors of a circuit can be disconnected from their source of supply.

Drum Controller. Electrical contacts made on surface of rotating cylinder or sector; contacts made also by operation of a rotating cam.

Drum Switch. A drum switch is a switch having electrical connecting parts in the form of fingers held by spring pressure against contact segments or surfaces on the periphery of a rotating cylinder or sector.

Duty. Specific controller functions. Continuous Duty: Constant load, indefinite long time period. Short Time Duty: Constant load, short or specified time period. Intermittent Duty: Varying load, alternate intervals, specified time periods. Periodic Duty. Intermittent duty with recurring load conditions. Varying Duty. Varying loads, varying time intervals, wide variations.

Dynamic Braking. Dynamic braking is accomplished by using the motor as a generator, taking it off the line and applying an energy dissipating resistor to the armature.

Eddy Currents. Circular induced currents contrary to the main currents; a loss of energy that shows up in the form of heat.

Electrical Interlocking. Electrical interlocking is accomplished by control circuits in which the contacts in one circuit control another circuit.

Electric Controller. An electric controller is a device, or group of devices, which governs, in some predetermined manner, the electric power delivered to the apparatus to which it is connected.

Electronic Control. Control system using gas and/or vacuum tubes or solid-state devices.

Enclosure. Mechanical, electrical and environmental protection for control devices.

Eutectic Alloy. Metal with low and sharp melting point; used in thermal overload relays; converts from a solid to a liquid state at a specific temperature; commonly called *solder pot*.

Feeder. A feeder is the circuit conductors between the service equipment, or the generator switchboard of an isolated plant, and the branch circuit overcurrent device.

Feeler Gauge. A precision instrument with blades in thicknesses of thousandths of an inch for measuring clearances.

Frequency. Number of complete variations made by an alternating current per second; expressed in hertz.

Full Load Torque (of a motor). The torque necessary to produce the rated horsepower of a motor at full load speed.

Full Voltage Control. (Across-the-line) Connects equipment directly to the line supply on starting.

Fuse. An overcurrent protective device with a fusible member which is heated directly and destroyed by the current passing through it to open a circuit.

General Purpose Motor. Any open motor having a continuous 40C rating and designed, listed, and offered, in standard ratings with standard operating characteristics and mechanical construction for use under usual service conditions without restrictions to a particular application or type of application. (NEMA)

High Voltage Control. Formerly, all control above 600 volts. Now, all control above 5000 volts. See Medium Voltage for 600- to 5000-volt equipment.

Horsepower. Measure of the time rate of doing work (working rate).

Instantaneous. Instantaneous is a qualifying term indicating that no delay is purposely introduced in the action of the device.

Interlock. To interrelate with other controllers; an auxiliary contact. A device is connected in such a way that the motion of one part is held back by another part.

Inverse Time. Inverse time is a qualifying term indicating that a delayed action is introduced purposely. This delay decreases as the operating force increases.

Jogging (Inching). Momentary operations; the quickly repeated closure of the circuit to start a motor from rest for the purpose of accomplishing small movements of the driven machine.

Limit Switch. A limit switch is a mechanically operated device which stops a motor from revolving or reverses it when certain limits have been reached.

Load Center. Service entrance; controls distribution; provides protection of power; generally of the circuit breaker type.

Local Control. Control function, initiation, or change accomplished at the same location as the electric controller.

Locked Rotor Current (of a motor). The steady-state current taken from the line with the rotor locked (stopped) and with the rated voltage and frequency applied to the motor.

Locked Rotor Torque (of a motor). The minimum torque that a motor will develop at rest for all angular positions of the rotor with the rated voltage applied at a rated frequency. (ASA)

Lockout. A mechanical device which may be set to prevent the operation of a pushbutton.

Low Voltage Protection (LVP). Magnetic control only; nonautomatic restarting; three-wire control; power failure disconnects service; power restored by manual restart.

Low Voltage Release (LVR). Magnetic control; automatic restarting; two-wire control; power failure disconnects service; when power is restored, the controller automatically restarts motor.

Magnet Brake. A magnet brake is a friction brake controlled by electromagnetic means.

Magnetic Contactor. A contactor which is operated electromechanically.

Magnetic Controller. An electric controller; device functions operated by electromagnets.

Maintaining Contact. A small control contact used to keep a coil energized, actuated by the same coil usually. Holding contact; Pallet switch.

Manual Controller. An electric controller; device functions operated by mechanical means or manually.

Master Switch. A switch to operate contactors, relays, or other remotely-controlled electrical devices.

Medium Voltage Control. Formerly known as High Voltage; includes 600- to 5000-volt apparatus; air break or oil-immersed main contactors; high interrupting capacity fuses; 150,000 kva at 2300 volts; 250,000 kva at 4000-5000 volts.

Motor. Device for converting electrical energy to mechanical work through rotary motion; rated in horsepower.

Motor Circuit Switch. Motor branch circuit switch rated in horsepower; capable of interrupting overload motor current.

Motor-Driven Timer. A device in which a small pilot motor causes contacts to close after a predetermined. time.

Multispeed Starter. An electric controller with two or more speeds; reversing or nonreversing; full or reduced voltage starting.

NEMA. National Electrical Manufacturers Association.

NEMA Size. Electric controller device rating; specific standards for horsepower, voltage, current, and interrupting characteristics.

Nonautomatic Controller. Requires direct operation to perform function; not necessarily a manual controller.

Nonreversing. Operation in one direction only.

Normally Open and Normally Closed. The terms *normally open* and *normally closed* when applied to a magnetically-operated switching device, such as a contactor or relay, or to the contacts of these devices, signify the position taken when the operating magnet is deenergized. These terms apply only to non-latching types of devices.

Overload Protection. Overload protection is the result of a device which operates on excessive current, but not necessarily on short circuit, to cause and maintain the interruption of current flow to the device governed. NOTE: Operating overload means a current that is not in excess of six times the rated current for alternating-current motors, not in excess of four times the rated current for direct-current motors.

Overload Relay. Running overcurrent protection; operates on excessive current; not necessarily protection for short circuit; causes and maintains interruption of device from power supply. Overload Relay Heater Coil: Coil used in thermal overload relays; provides heat to melt eutectic alloy. Overload Relay Reset: Pushbutton used to reset thermal overload relay after relay has operated.

Panelboard. Panel, group of panels, or units; an assembly which mounts in a single panel; includes buses, with or without switches and/or automatic overcurrent protective devices; provides control of light, heat, power circuits; placed in or against wall or partition; accessible from front only.

Permanent-Split Capacitor Motor. A single-phase induction motor similar to the capacitor start motor except that it uses the same capacitance which remains in the circuit for both starting and running. (NEMA)

Permeability. The ease with which a material will conduct magnetic lines of force.

Phase. Relation of current to voltage at a particular time in an ac circuit. Single Phase: A single voltage and current in the supply. Three Phase: Three electrically related (120-degree electrical separation) single-phase supplies.

Phase-Failure Protection. Phase-failure protection is provided by a device which operates when the power fails in one wire of a polyphase circuit to cause and maintain the interruption of power in all the wires of the circuit.

Phase-Reversal Protection. Phase-reversal protection is provided by a device which operates when the phase rotation in a polyphase circuit reverses to cause and maintain the interruption of power in all the wires of the circuit.

Phase Rotation Relay. A phase rotation relay is a relay which functions in accordance with the direction of phase rotation.

Pilot Device. Directs operation of another device. Float Switch: A pilot device which responds to liquid levels. Foot Switch: A pilot device operated by the foot of an operator. Limit Switch: A pilot device operated by the motion of a power-driven machine; alters the electrical circuit with the machine or equipment. Pressure Switch: A pilot device operated in response to pressure levels. Temperature Switch: A pilot device operated in response to temperature values.

Plugging. Braking by reversing the line voltage or phase sequence; motor develops retarding force.

Pole. The north or south magnetic end of a magnet; a terminal of a switch.

Potentiometer. A variable resistor with two outside fixed terminals and one terminal on the center movable arm.

Pull-up Torque (of alternating-current motor). The minimum torque developed by the motor during the period of acceleration from rest to the speed at which breakdown occurs. (ASA)

Pushbutton. A master switch; manually operable plunger or button for an actuating device; assembled into pushbutton stations.

Relay. Operated by a change in one electrical circuit to control a device in the same circuit or another circuit; rated in amperes; used in control circuits.

Remote Control. Controls the function initiation or change of an electrical device from some remote point.

Remote Control Circuit. Any electrical circuit which controls any other circuit through a relay or an equivalent device.

Residual Magnetism. The retained or small amount of remaining magnetism in the magnetic material of an electromagnet after the current flow has stopped.

Resistance. Resistance is the opposition offered by a substance or body to the passage through it of an electric current; resistance converts electrical energy into heat; resistance is the reciprocal of conductance.

Resistor. A resistor is a device used primarily because it possesses the property of electrical resistance. A resistor is used in electrical circuits for purposes of operation, protection, or control; commonly consists of an aggregation of units.

> *Starting Resistors:* Used to accelerate a motor from rest to its normal running speed without damage to the motor and connected load from excessive currents and torques, or without drawing undesirable inrush current from the power system.

> *Armature Regulating Resistors:* Used to regulate the speed or torque of a loaded motor by resistance in the armature or power circuit.

> *Dynamic Braking Resistors:* Used to control the current and dissipate the energy when a motor is decelerated by making it act as a generator to convert its mechanical energy to electrical energy and then to heat in the resistor.

> *Field Discharge Resistors:* Used to limit the value of voltage which appears at the terminals of a motor field (or any highly inductive circuit) when the circuit is opened.

> *Plugging Resistors:* Used to control the current and torque of a motor when deceleration is forced by electrically reversing the motor while it is still running in the forward direction.

Rheostat. A resistor that can be adjusted to vary its resistance without opening the circuit in which it may be connected.

Safety Switch. Enclosed manually-operated disconnecting switch; horsepower and current rated; disconnects all power lines.

Selector Switch. A master switch that is manually operated; rotating motion for actuating device; assembled into pushbutton master stations.

Semiautomatic Starter. Part of the operation of this type of starter is nonautomatic while selected portions are automatically controlled.

Semimagnetic Control. An electric controller whose functions are partly controlled by electromagnets.

Sensing Device. A pilot device that measures, compares, or recognizes a change or variation in the system which it is monitoring; provides a controlled signal to operate or control other devices.

Service. The conductors and equipment necessary to deliver energy from the electrical supply system to the premises served.

Service Equipment. Necessary equipment, circuit breakers or switches and fuses, with accessories mounted near the entry of the electrical supply; constitutes the main control or cutoff for supply.

Service Factor (of a general-purpose motor). An allowable overload; the amount of allowable overload is indicated by a multiplier which, when applied to a normal horsepower rating, indicates the permissible loading.

Shaded Pole Motor. A single-phase induction motor provided with auxiliary short-circuited winding or windings displaced in magnetic position from the main winding. (NEMA)

Slip. Difference between rotor rpm and the rotating magnetic filed of an ac motor.

Solenoid. A tubular, current-carrying coil that provides magnetic action to perform various work functions.

Solenoid-and-plunger. A solenoid-and-plunger is a solenoid provided with a bar of soft iron or steel called a plunger.

Solder Pot. See Eutectic Alloy.

Special-Purpose Motor. A motor with special operating characteristics or special mechanical construction, or both, designed for a particular application and not falling within the definition of a general-purpose or definite-purpose motor. (NEMA)

Split-phase. A single-phase induction motor with auxiliary winding, displaced in magnetic position from, and connected in parallel with, the main winding. (NEMA)

Starter. A starter is a controller designed for accelerating a motor to normal speed in one direction of rotation. NOTE: A device designed for starting a motor in either direction of rotation includes the additional function of reversing and should be designated as a controller.

Static Control. Control system in which solid-state devices perform the functions.

Switch. A switch is a device for making, breaking, or changing the connections in an electric circuit.

Switchboard. A large, single panel with a frame or assembly of panels; devices may be mounted on the face of the panels, on the back or both; contains switches, overcurrent, or protective devices; instruments accessible from the rear and front; not installed in wall-type cabinets (see Panelboard).

Synchronous Speed. Motor rotor and ac rotating magnetic field in step or unison.

Tachometer Generator. Used for counting revolutions per minute. Electrical magnitude or impulses are calibrated with a dial gage reading in rpm.

Temperature Relay. A temperature relay is a relay which functions at a predetermined temperature in the apparatus protected. This relay is intended to protect some other apparatus such as a motor or controller and does not necessarily protect itself.

Thermal Protector (as applied to motors). An inherent overheating protective device which is responsive to motor current and temperature. When properly applied to a motor, this device protects the motor against dangerous overheating due to overload or failure to start.

Time Limit. See Definite Time.

Timer. A pilot device that is also considered a timing relay; provides adjustable time period to perform function; motor driven; solenoid actuated; electronic.

Torque. The torque of a motor is the twisting or turning force which tends to produce rotation.

Transducer. A device that transforms power from one system to power of a second system: example, heat to electrical.

Transformer. Converts voltages for use in power transmission and operation of control devices; an electro-magnetic device.

Trip Free. Refers to a circuit breaker which cannot be held in the on position by the handle on a sustained overload.

Undervoltage Protection. Undervoltage protection is the result when a device operates on the reduction or failure of voltage to cause and maintain the interruption of power to the main circuit.

Undervoltage Release. Undervoltage release is the result when a device operates on the reduction or failure of voltage to cause the interruption of power to the main circuit but does not prevent the reestablishment of the main circuit on the return of voltage.

Voltage Relay. A voltage relay is a relay which functions at a predetermined value of voltage. A voltage relay may be either an overvoltage or an undervoltage relay.

Appendix-General Information

MOTOR TYPES

Torque increases with load

Dc Shunt Motor. Main field winding is designed for parallel connection to the armature; stationary field; rotating armature with commutator; has a no-load speed; full speed at full load is less than no-load speed; torque increases directly with load.

Excellent Starting Torque

Dc Series Motor. Main field winding is designed for series connection to the armature; stationary field; rotating armature with commutator; does not have a no-load speed; requires solid direct connection to the load to prevent runaway at no-load; speed decreases rapidly with increase in load; torque increases as square of armature current; main motor for crane hoists; excellent starting torque.

Good Starting Torque

Dc Compound Motor. Main field both shunt (parallel) and series; stationary fields; rotating armature with commutator; combination shunt and series fields produce characteristics between straight shunt or series dc motor; good starting torque; main motor for dc driven machinery (mills or presses).

Good Starting torque

Ac Squirrel Cage Motor. Single or three phase; single phase requires a starting winding; three phase, self starting; stationary stator winding; no electrical connection to short circuited rotor; torque produced from magnetic reaction of stator and rotor fields; speed a function of supply frequency and number of electrical poles wound on stator; considered as constant speed even though speed decreases slightly with increased load; good starting torque; high inrush currents during starting on full voltage; rugged construction; easily serviced and maintained; high efficiency; good running power factor when delivering full load; requires motor control for stator windings only.

Ac Wound-Rotor Induction Motor. Characteristics similar to squirrel cage motor; stationary stator winding; rotor windings terminate on slip rings; external addition of resistance to rotor circuit for speed control; good starting torque; high inrush current during starting on full voltage; low efficiency when resistor is inserted in rotor windings; good running power factor; requires motor controls for stator and rotor circuits.

Poor Starting Torque

Ac Synchronous Motor. Stationary ac stator windings; rotating dc field winding; no starting torque unless motor has starting winding; generally poor starting torque; constant speed when motor up to speed and dc field winding energized; can provide power factor correction with proper dc field excitation; requires special motor control for both ac and dc windings to prevent the dc field winding from being energized until a specified percent of running speed has been obtained.

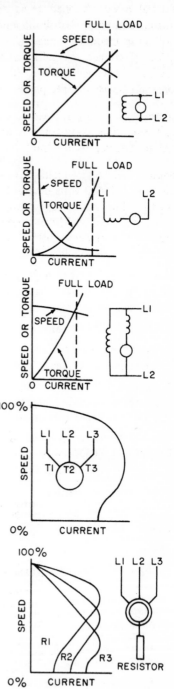

POWER SUPPLIES

All electrical power supplied as ac or dc; primarily ac; generation, transmission, and some distribution of power at high voltage (above 5000 volts) or medium voltage (600 to 5000 volts); most power distribution of voltage for industrial and residential use is 600 volts and under; ac power generally at 60-hertz frequency; ac distribution at use location single or three phase.

Single-phase. Two wire, 120 volts, one line grounded; 120/240 volts, three-wire centerline grounded; residential distribution, lighting, heat, fractional horsepower motors and business machines.

Three-phase. Three-wire delta 220/440 volts, 550 volts; four-wire wye 208/440, 277/480 volts neutral line grounded; primary industrial power distribution; main motor drives, integral horsepower motors, lighting, heating, fractional horsepower motors, and business machines; used as three-phase or single-phase power supply.

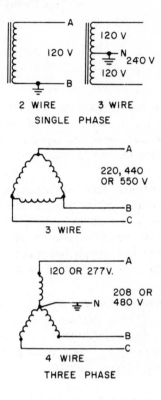

MOTOR CIRCUIT ELEMENTS

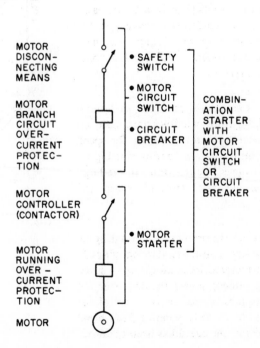

AIDS TO SERVICING ELECTRIC CONTROL EQUIPMENT

TABLE I* INDUSTRIAL CONTROLS AND INDEX TO MAINTENANCE SUGGESTIONS

Type of Control	Control Symptoms (See Table II)

Electrically Operated Devices

I. Magnetically operated Device

A. Contactors

  1. Ac
    a. Ac operated   (1-12 inc.)
    b. Dc operated   (5-13 inc.)
  2. Dc   (5-13 inc.)

B. Relays

  1. Ac
    a. Simple magnetic   (1-12 inc.)
    b. Timing (mechan-   (14,15)
      ical escapement type)
    c. Overcurrent
      (1) Instantaneous   (16,17)
      (2) Inverse-time   (16-19 inc.)

  2. Dc
    a. Simple magnetic   (5-13 inc.)
    b. Timing
      (1) Mechanical   (14,15)
        escapement
      (2) Decay-of-flux   (7-13 inc., 20,21)
      (3) Capacitor   (7-13 inc., 22,23)
    c. Overcurrent
      (1) Instantaneous   (16, 17)
      (2) Inverse-time   (16-19 inc.)

C. Solenoids

  1. Ac operated   (24-30 inc.)
  2. Dc operated   (28-31 inc.)

D. Valves

  1. Ac operated   (27,32-35 inc.)
  2. Dc operated   (34-36 inc.)

E. Brakes

  1. Shoe brake
    a. Ac operated   (24-30 inc., 37-41 inc.)
    b. Dc operated   (28-31 inc., 38-41 inc.)
  2. Disc brake
    a. Ac operated   (21,26,37-41 inc.)
    b. Dc operated   (12,31,38-41 inc.)

II. Motor-operated device

A. Relays

  1. Timing   (11,15,42,43)
  2. Induction disc or cup type   (11,44)

B. Brakes
  1. Thrustor operated   (38-41 inc., 45)

C. Thrustors   (45)

D. Valves
  1. Thrustor operated   (35,45,46)
  2. Geared   (35, 46)

E. Rheostats   (11,47,48)

III. Thermally Operated Device

A. Relays   (11, 49-52 inc.)

B. Thermostats
  1. Bimetallic   (11,53)
  2. Expanding fluid   (11,52,54,55)
    (bulb and bellows)

IV. Static Accessories

A. Resistors   (11, 56-58 inc.)

B. Rectifiers (dry type)   (59)

C. Capacitors   (6A)

D. Transformers   (61, 62)

E. Fuses   (63, 64)

Mechanically Operated Devices

I. Manually Operated Devices

A. Master switches   (11,52,65)

B. Drum controllers   (8,11,52,65-67)

C. Pushbuttons   (11,52,65,68,69)

D. Selector switches   (11,52,65,68,69)

E. Knife switches   (70)

F. Manual starters
  1. Full voltage (small sizes)   (8,11,49,50,65)
  2. Reduced voltage   (8,11,49,50,65)

G. Rheostats   (11, 47, 48)

H. Manual contactors   (8, 11, 65)

II. Otherwise Mechanically Operated Devices

A. Limit switches   (8, 11, 52, 71)

B. Speed-sensitive switches   (11, 14, 52, 71)

C. Float switches   (11, 14, 52, 65)

D. Flow switches   (11, 14, 52, 65, 72)

E. Pressure switches   (11, 14, 52, 65, 72)

Electrical Control Troubles and Their Causes by J. H. Hopper, Industrial Control Engineering Dept., General Electric Co.

TABLE II* Industrial Control Troubleshooting Tips

Control Symptoms	Possible Causes and Things to Investigate	Control Symptoms	Possible Causes and Things to Investigate
1. Noisy magnet	Broken pole shader; magnet faces not true as result of wear or mounting strains; dirt on magnet.	13. Coil failure	Overvoltage; high ambient; wrong coil; moisture; corrosive atmosphere; intermittent coil energized continuously; holding resistor not cut in.
2. Broken pole shader	Heavy slamming caused by overvoltage, weak tip pressure, wrong coil.	14. Sticking	Dirt; worn parts; improper adjustment; corrosion; mechanical binding.
3. Coil failure	Moisture; overvoltage; high ambient; failure of magnet to seal in on pickup; too rapid duty cycle; corrosive atmosphere; chattering of magnet; metallic dust.	15. Mechanical wear	Abrasive dust; improper application (in general not suited for continuous cycling).
		16. Low trip	Wrong coil; assembled wrong.
4. Wear on magnet	Overvoltage; broken pole shader; wrong coil; weak tip pressure; chattering.	17. High trip	Mechanical binding; wrong or shorted coil; assembled wrong.
5. Blowout coil overheats	Overcurrent; wrong size of coil; loose connections on stud or tip; tip heating (see Table I); excess frequency.	18. Fast trip	Fluid out or too light; vent too large; high temperature; wrong heaters.
		19. Slow trip	Fluid too heavy; vent too small; mechanical binding; dirt; low temperature; wrong heaters.
6. Pitted, worn, or broken arc chutes	Abnormal interrupting duty (inductive load); excessive vibration or shock; moisture; improper assembly; rough handling.	20. Too-short time	Dirt in air gap; shim too thick; too much spring and tip pressure; misalignment.
7. Failure to pick up	Low voltage; coil open; mechanical binding; no voltage; wrong coil; shorted turns; excessive magnet gap.	21. Too-long time	Shim too thin; weak spring and tip pressure (use brass screws for steel backed shims); gummy substance on magnet faces.
8. Contact-tip troubles	See Table I.	22. Too-short time	Same as 20, plus not enough capacitance; not enough resistance.
9. Broken flexible shunt	Large number of operations; improper installation (check instructions); extreme corrosive conditions; burned from arcing.	23. Too-long time	Same as 21, plus too much capacitance; too much resistance.
10. Failure to drop out	Mechanical binding; gummy substance on magnet faces; air gap in magnet destroyed; contact tip welding; weak tip pressure; voltage not removed.	24. Noisy magnet	Same as 1, plus low voltage; mechanical overload; load out of alignment.
		25. Broken pole shader	Heavy slamming caused by overvoltage; mechanical underload; wrong coil; low frequency.
11. Insulation failure	Moisture; acid fumes; overheating; accumulation of dirt on surfaces; voltage surges; short circuits.	26. Coil failure	Same as 3, plus mechanical overload (can't pick up); mechanical underload (slam); mechanical failure.
12. Various mechanical failures	Overvoltage; heavy slamming (see 2) chattering; abrasive dust.	27. Wear on magnet	Overvoltage; underload; broken pole shader; wrong coil; chattering; load out of alignment.

Electrical Control Troubles and Their Causes by J. H. Hopper, Industrial Control Engineering Dept., General Electric Co.

TABLE II* Industrial Control Troubleshooting Tips

Control Symptoms	Possible Causes and Things to Investigate	Control Symptoms	Possible Causes and Things to Investigate
28. Failure to pick up	Same as 7, plus mechanical overload	43. Failure to reset	Mechanical binding; worn parts; dirt.
29. Failure to drop out	Mechanical binding; gummy substance on magnet faces; air gap in magnet destroyed; voltage not removed; contacts welded.	44. Failure to operate properly	Coils connected wrong; wrong coils; mechanical binding.
30. Miscellaneous mechanical failures	Same as 12, plus underload.	45. Short life of thrustor	Abrasive dust; dirty oil; water in tank.
31. Coil failure	Same as 13, plus mechanical failure of coil.	46. Failure to open or close	Corrosion; scale; dirt; mechanical binding; damaged motor; no voltage.
32. Noise	Same as 1, plus water hammer.	47. Wear on segments or shoes	Abrasive dust; very heavy duty; no lubrication (special materials available.)
33. Coil failure	Same as 3, plus handling fluid above rated temperature.	48. Resistor failure	Overcurrent; moisture; corrosive atmospheres.
34. Failure to open or close	Similar to 28,29, plus corrosion; scale; dirt; operating above rated pressure.	49. Failure to trip (motor burnout)	Heater incorrectly sized; mechanical binding; relay previously damaged by short circuit current; dirt; corrosion; motor and relay in different ambient temperatures.
35. Leaks and mech. failure	Worn seat; solid matter in seat (check strainer ahead of valve).	50. Failure to reset	Broken mechanism; corrosion; dirt; worn parts; resetting too soon.
36. Coil failure	Same as 13, plus handling fluid above rated temperature.	51. Trips too low	Wrong heater; relay in high ambient (check temperature of motor).
37. Noisy magnet	Same as 2, plus improper adjustment (too much spring pressure or incorrect lever ratio).	52. Burning and welding of control contacts and shunts	Short circuits on control circuits with too large protecting fuses; severe vibration; dirt; oxidation.
38. Excess wear or friction	Abrasive dust; heavy-duty; high inertia load; excess temperature; scarred wheels.		
39. Failure to hold load	Worn parts; out of adjustment; misapplication; failure to use recommended substitute parts.	53. Arcing and burning of contacts	Misapplied; should handle very little current and have sealing circuit.
40. Failure to set	Improper adjustment; mechanical binding; coil not deenergized; worn parts.	54. Bellows distorted	Mechanical binding; temperature allowed to overshoot.
41. Failure to release	Improper adjustment; coil not energized; mechanical binding; low voltage or current; coil open; shorted turns.	55. Bulb distorted	Liquid frozen in capillary tube.
		56. Overheating	Used above rating; running on starting resistor.
42. Failure to time out	Mechanical binding; worn parts; motor damaged; no voltage on motor; dirt.	57. Corrosion	Excess moisture; salt air; acid fumes.

Electrical Control Troubles and Their Causes by J. H. Hopper, Industrial Control Engineering Dept., General Electric Co.

TABLE II* Industrial Control Troubleshooting Tips

Control Symptoms	Possible Causes and Things to Investigate	Control Symptoms	Possible Causes and Things to Investigate
58. Breakage, distortion and wear	Overheating; mechanical abuse; severe vibration; shock.	67. Flashover	Jogging; short circuits; handling too large motor; moisture; acid fumes; gases; dirt.
59. Breakdown	High temperature; moisture; over-current; overvoltage; corrosive atmospheres; mechanical damage.	68. Failure to break arc	Too much current; too much voltage; (usually dc); misapplication; too much inductance.
60. Breakdown	Overload; overvoltage; ac on dc capacitor; moisture; high temperature; mechanical damage; continuous voltage on intermittent types.	69. Failure to make contact	Mechanical damage; dirt; corrosion; wear allowance gone.
61. Overheating	Overload; overvoltage; intermittent-rated device operated too long.	70. Heating	Overcurrent; loose connection; spring clips loose or annealed; oxidation; corrosion.
62. Insulation failure	Overheating; overvoltage; voltage surges; moisture; mechanical damage.	71. Mechanical failure	Same as 65, plus excessive operating speed.
63. Premature blowing	Extra heating from outside; copper oxide on ferrules and clips (plated ferrules and clips are available); high ambient.	72. Leaks	Corrosion; mechanical damage; excessive pressure.
64. Too slow blowing	Wrong sized fuse for application.		
65. Mechanical failures	Abrasive dust and dirt; misapplication; mechanical damage.		
66. Short contact life	Jogging; handling abnormal currents; lack of lubrication where recommended; abrasive dirt.		

Electrical Control Troubles and Their Causes by J. H. Hopper, Industrial Control Engineering Dept., General Electric Co.

ACKNOWLEDGMENTS

Publication Director
Alan N. Knofla

Source Editor
Marjorie A. Bruce
Supervising Editor
Technical Education Division

Revision
Walter N. Alerich
Coordinator of Instruction
Electrical-Mechanical Dept.
Los Angeles Trade-Technical College
Los Angeles, California

Director of Manufacturing and Production
Frederick Sharer

Production Specialists
Lee St. Onge
Jean Le Morta
Betty Michelfelder
Patti Barosi
Sharon Lynch

Illustration
Anthony Canabush   George Dowse
Michael Kokernak   Chris Carline

The author wishes to express his appreciation to the following companies for their assistance in illustrating this text.

Warner Electric Brake and Clutch Co.
Figure 58-3

Electric Machinery Manufacturing Co.
Figures 1-1, 1-2, 1-6, 1-7, 1-9, 30-1, 34-1, 34-3, 37-4, 37-5, 58-7

U.S. Electrical Motors
Figures 1-3, 23-1, 56-2, 57-2

Square D Company
Figures 1-4, 1-10, 1-11, 2-1, 2-6, 2-7, 3-1, 3-3, 3-4, 3-5, 3-8, 3-14, 3-15, 3-16,
5-1, 5-3, 5-4, 5-8, 6-1, 7-1, 10-1, 12-2, 20-1, 22-1, 23-3, 23-7, 25-1,
27-1, 47-5, 50-1, 54-1, 59-1

Cutler-Hammer Inc.
Figures 1-5, 1-8, 30-2, 41-2, 41-4, 52-1, 52-3, 52-4

Allen-Bradley Co.
Figures 3-6, 4-1, 4-2, 4-3, 5-2, 6-2, 6-4, 14-1, 15-1, 21-1, 23-8, 24-1, 37-1,
37-2, 37-3, 38-1, 38-3, 51-1

General Electric Co.
Figures 6-6, 6-10, 47-1, 47-3

Automatic Switch Co.
Figure 12-1

A

Accelerating relays, 114-115
Ac combination starters, 24-26
Ac reduced voltage starters, 82-90
    autotransformer starters, 91-93
    part winding motor starters, 94-96
    for star-delta motors, 97-100
Across-the-line starting, 143-144
Ac squirrel cage motor, 216
Ac synchronous motor, 216
Ac wound-rotor induction motor, 216
Adjustable speed motors, 9
Antiplugging protection, 177-178
Automatic acceleration
    with reversing control, 122
    using frequency relays, 122-124
    for wound rotors motors, 121-124
Automatic control, 5-6
    current limiting acceleration, 11
    hands-off, 70
    time delay acceleration, 11
Automatic sequence control, 79
Automatic speed control, 125
Automatic starters
    for star-delta motors, 97-100
    summary of, 139-140
Autotransformer starters, 82, 91-93
Auxiliary contact interlock, 75-77

B

Bimetallic overload relays, 18-19
Blowout coils, 31
Blowout magnet, 31-32
Brakes *See* Dc magnetic brakes; Electric brakes;
    Magnetic disc brakes
Braking *See* Dynamic braking; Electric braking

C

Capacitor time limit relay, 39-40
Capacitor timing, 166
Capacitor timing starter, 166-167
Class 1, Group D enclosures, 26
Closed transition starting, 99
Clutch *See* Magnetic clutch; Multipleface clutch;
    Single-face clutch
Combination starter, 24-26
Compelling relays, 111-114
Compensation time *See* Current limiting acceleration
Compensator starting *See* Autotransformer starting
Contactors, 30-32
    mechanically held, 32
Control circuit
    separate control, 69

    three-wire controls, 68-69
    two-wire controls, 67
Controller
    operator, safety of, 3
    protection from damage, 3, 9-10
    purpose of, 2-4
    reversing, 3
    running, 3
    speed controller, 3
    starting, 3
    starting requirements, maintenance of, 4
    stopping, 3
    for two-speed two winding motors, 102-104
Control pilot devices
    contactors, 30-32
    pushbutton control, 27-28
    relays, 29-30, 32
    timing relays, 34-41
Constant speed motors, 8
Control relays, 29-30, 32
    jogging, used in, 169-170
    mechanical latch relay, 141
    thermostat relay, 141-142
Counter EMF controller, 150
Current limiting acceleration, 11

D

Dashpot, 160
Dashpot motor control, 160-163
Dashpot principle, 19
DC compound motor, 216
Dc magnetic brakes, 180-182
Dc series magnetic lockout contactor, 38
Dc series motor, 216
Dc series relay, 37-38
Dc shunt motor, 216
Dc variable speed control motor drives, 206-209
Decelerating relays, 114-115
Deceleration, methods of
    dynamic braking, 183-187
    electric brakes, 179-182, 188-191
    jogging control circuits, 168-171
    plugging, 172-178
Diagrams, 56-65
    control and power connections, 59-60
    motor nameplates and, 62-63
    single-phase, dual voltage motor connections, 64-65
Direct-current controllers, 141
Directly-coupled drive installation, 192-194
Disc brakes *See* Magnetic disc brakes
Drum controller, 4
Drum switch diagramming, 60-61

Dustight enclosures, 26
Dynamic braking, 183-187

E

Eddy currents, 22
Electrical interlock, 6
Electrical symbols, 53-55
Electric brakes, 179-182
Electric braking, 188-191
Electric motor control, 1-11
Elementary diagrams *See* Line diagrams

F

Faceplate control, 4
Field accelerating relay, 160
Field contractor, 135
Field failure relay, 160-162
Float switch, 5, 44
Flow switches, 45-46
Fluid dashpot timing relays, 34-35
Four-speed two winding motor controller, 111-114
Four-terminal dc faceplate starters, 149
Four way solenoid valves, 50-51
Fractional horsepower manual motor starters
    automatic operation and, 12-15
    thermal overload protection, 14-15
Friction brakes *See* Electric brakes
Full field relay *See* Field accelerating relay
Full voltage starting, 82

G

Gear motors, 197-199
General-purpose enclosures, 25

H

Hand-off automatic controls, 70

I

Impedance starting, 83
Inching *See* Jogging
Inching control circuits *See* Jogging control circuits
Instantaneous trip current relays, 20
Interlocking methods
    auxiliary, 75-77
    mechanical, 75
    pushbutton, 75
    for reversing control, 74-77

J

Jogging
    control relay, use of, 167-170
    defined, 168
    selector switch, use of, 170-171
Jogging control circuits, 168-171

L

Limit switch, 6, 47
Line diagrams, 56-58

Line voltage magnetic starters *See* Magnetic line
    voltage starters
Locked rotor currents, 85
Low-voltage protection *See* Three-wire control
Low-voltage release *See* Two-wire control

M

Magnetic blowout, 30-31
Magnetic brakes *See* Electric brakes
Magnetic clutch, 202-204
Magnetic control, 16
Magnetic disc brakes, 180
Magnetic drives, 204-205
Magnetic line voltage starters, 17-21
Magnetic overload relays, 19
Magnetic relays, 29-30
Magnetic time limit control, 152-153
Magnetic time limit relay, 38-39
Manual control, 4
Manual pushbutton line voltage starters, 15
Manual speed control, 116-118
Manual starter *See also* Fractional manual horsepower
Mechanical brakes *See* Electric brakes
Mechanical held contactors, 32
Mechanical interlock, 6, 75
Mechanical latch relay, 141
Mechanically held relays, 32
Mechanical protection, 10
Melting alloy thermal units, 18
Motor
    revolving field, 83-87
    speed control, 8-9
Motor control
    defined, 2
    protective features, 9-10
    purpose of, 2-4
    starting and stopping, 6-8
Motor driven timers, 36
Motor drives
    dc variable speed control, 206-209
    direct drives, 192-194
    gear motors, 197-199
    pulley drives, 194-196
    variable frequency drives, 200-201
Motor nameplate data, 62-63
Multicircuit time limit dashpot acceleration, 162-163
Multiple-face clutch, 203
Multiple pushbutton stations, 72
Multispeed motors, 9

N

No-load rotor speed, 84
No-voltage protection *See* Three-wire control

No-voltage release *See* Two-wire control

O

Open field protection, 9
Open-phase protection, 9
Out-of-step relay, 135-136
Overload protection, 9
    magnetic line voltage starters and, 17-21
    part winding motor starters, 95
Overspeed protection, 10
Overtravel protection, 10

P

Pairs of poles, 101
Part winding motor starters, 94-96
Part winding starting, 83
Phase failure relays, 48
Pilot motor-driven timer controller, 164-165
Plugging, 172-178
Plugging switches, 172-174, 175-177
Pneumatic timers, 35-36
Polarized field frequency relay, 136-139
Pressure regulators, 43
Pressure switch, 5, 43
Primary resistance starting, 82
Primary resistor-type starters, 87-90
Protective enclosures, 24-26
Pulley drives, 194-195
Pushbutton control, 27-28
Pushbutton interlock, 75
Pushbutton speed selection, 119-120
Pushbutton synchronizing, 131-132

R

Reduced voltage starting, 85-86
Regenerative braking, 183-185
Relays
    See also Timing relays
    compelling relays, 111-114
    control relays, 29-30, 32
    mechanically held, 32
    pneumatic timers, 35-36
Remote control, 5-6
Reversed current protection, 10
Reversed phase protection, 10
Reversing starter, 170
Revolving field, 83-87
Rotor, 83-84
Rotor control equipment
    field contractor, 135
    out-of-step relay, 135-136
    polarized field frequency relay, 136-139
Running protection *See* Overload protection

S

Safety switch, 4
Schematic diagram *See* Line diagram
Secondary resistor starters *See* Wound rotor motor controllers
Selector switch, 170-171
Selector switch diagramming, 60
Separate controls, 69
Sequence chart *See* Target tables
Sequence control, 78-79
Series lockout relay acceleration, 158-159
Series relay, 37-38
Series relay acceleration, 156-157
Series starting resistance, 145-146
Shaded pole principle, 22-24
Short circuit protection, 10
Silicone dashpot fluid, 34-35
Single-face clutch, 203
Slip, 84-85
Slip ring motors *See* Wound rotor motors
Solenoid, 21-22
Solenoid valves, 49-51
Squirrel cage induction motor, 101
Star-delta motors
    automatic starters for, 97-100
    operation of, 98-99
Star-delta starting, 83
Starter electromagnets, 21-22
Starters
    across-the-line, 143-144
    ac reduced voltage starters, 82-100
    autotransformer starters, 91-93
    compelling relays, 111-114
    four-terminal dc faceplate starters, 149
    magnetic line voltage starters, 16-26
    manual faceplate starters, 147-149
    operational sequences, 111
    part winding motor starters, 94-96
    primary resistor type, 82-90
    three-terminal dc faceplate starter, 147-149
Starting, 6-8
Starting methods, 82-83
Stopping, 6-8
Symbols, 53-55
Synchronous motor
    operation of, 127-129
    power factor correction by, 128-129
Synchronous motor control, 131
Synchronous motor starter, 135-140

T

Target tables, 60-61
Temperature switches, 52

Thermal overload protection, 14-15
Thermostat, 5, 52
Thermostat relay, 141-142
Three-phase multispeed controller
    for two-speed one winding motor, 105-109
    for two-speed two winding motor, 101-104
Three-wire control, 59, 68-69, 80
Time Clock, 5
Time delay, 23
Time delay acceleration, 11
Time-delay, low voltage release relay, 80-81
Timed semiautomatic synchronizing, 133-134
Time limit overload relays, 19-20
Timing range, 40
Timing relays
    capacitor time limit relay, 39-40
    dc series lockout, 38
    fluid dashpot, 34-35
    magnetic time limit relay, 38-39
    pneumatic timers, 35-36
    selections of, 40-41
Toggle switch, 4

Torque, control of, 85
Two-speed one-winding motor controller, 105-109
Two-speed separate winding motors, 101-104
Two-speed starter with reversing controls, 109
Two-way solenoid valves, 49-50
Two-wire control, 58, 67

V

Variable frequency drives, 200-201
Varying speed motors, 8
Voltage drop acceleration, 154-155

W

Watertight enclosures, 25
Wiring diagrams, 56-58
Wound motor controller, 116-118
Wound rotor motors
    automatic acceleration for, 121-124
    automatic speed control for, 125
    electric braking for, 190-191
    pushbutton speed selection for, 119-120

Z

Zero-speed switches *See* Plugging switches